W9-AVW-280

The
Homesteader's
Handbook to

RAISING SMALL LIVESTOCK

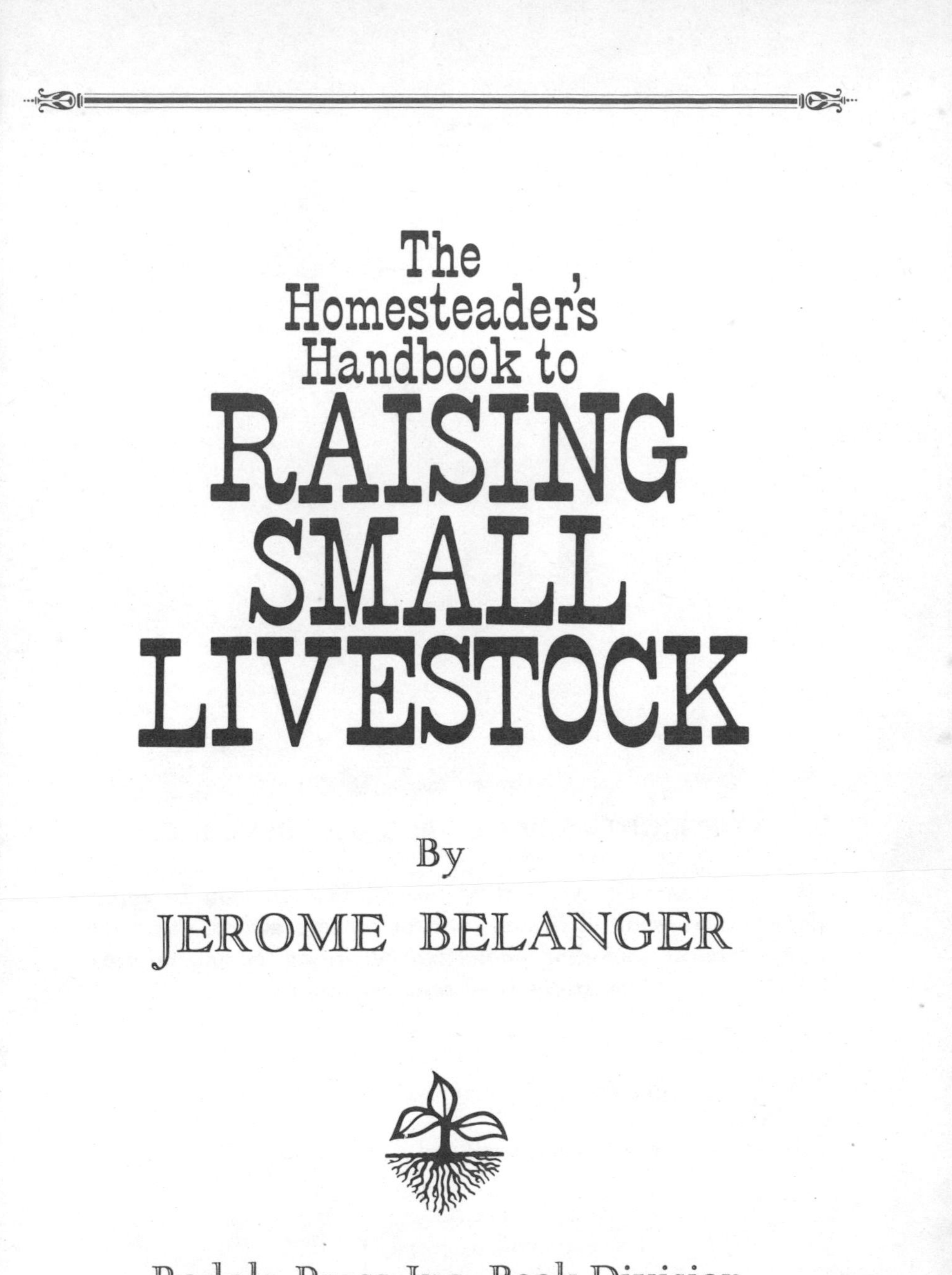

The Homesteader's Handbook to RAISING SMALL LIVESTOCK

By

JEROME BELANGER

Rodale Press Inc. Book Division
Emmaus, Pennsylvania 18049

COPYRIGHT © 1974 BY RODALE PRESS, INC.

All rights reserved. No part of this publication may be reproduced or transmitted in any form or by any means, electronic or mechanical, including photocopy, recording, or any information storage and retrieval system.

Standard Book Number 0-87857-075-6

Library of Congress Card Number 73-88254

Thirteenth Printing—September 1977

Printed in the United States of America

CONTENTS

INTRODUCTION

Like many Americans my age and older, my first experience with small stock was during World War II. We lived in town, but with meat and eggs rationed or unavailable, we raised rabbits and chickens and, of course, a victory garden.

Similar factors account for some of the interest in small stock today. Not only has the quality of supermarket food deteriorated in recent years, but there are indications that even the *quantity* we've become accustomed to might not be available. Just the price of food at the supermarket makes subsistence farming attractive to many people.

But it goes far beyond that. For many raisers of small stock, food production is almost of secondary importance. They claim benefits that range from strictly mundane to almost spiritual. Some are interested in bedding and manure for the organic garden. I actually know people who keep a few rabbits just for the manure, which they use on their worm beds. Others simply enjoy the antics and personalities of their animals, or the fun of showing them. Still others claim that raising animals gives them an increased respect for life itself.

Just the same, the one word that sums up the reason most people raise small stock today is *independence.* Freedom from the chemical companies, from price increases, from shortages. Being independent, in this day and age, is something you have to work at, and in keeping with our pioneer traditions, a state worth working for.

Even those staunch individualists, the farmers, aren't independent today in this sense.

Our family lives on a farm in southern Wisconsin. We raise over 1,000 hogs a year and a number of beef cattle. Most of our neighbors with similar size operations sell their hogs and cattle and take the money to town to buy eggs, milk, and all their other groceries. Many don't even have gardens.

We, on the other hand, have a huge garden, 150 chickens, ducks and geese and guineas, rabbits, sheep and goats, in addition to the hogs and beef cattle. We raise our own wheat and grind our own flour. We make cheese, can hundreds of quarts of fruit and vegetables, and do much of our own butchering.

Work? Of course. But there's a certain grim satisfaction in being "independent." Moreover, just because it's work doesn't mean it isn't fun!

* * * * * *

You'd think that anybody who'd write a book on small stock would be an expert on the subject. I did and I'm not.

There are no experts. People who have concentrated on even one species of animals for 20, 30, even 50 years, still learn something new every day. Far from being a handicap,

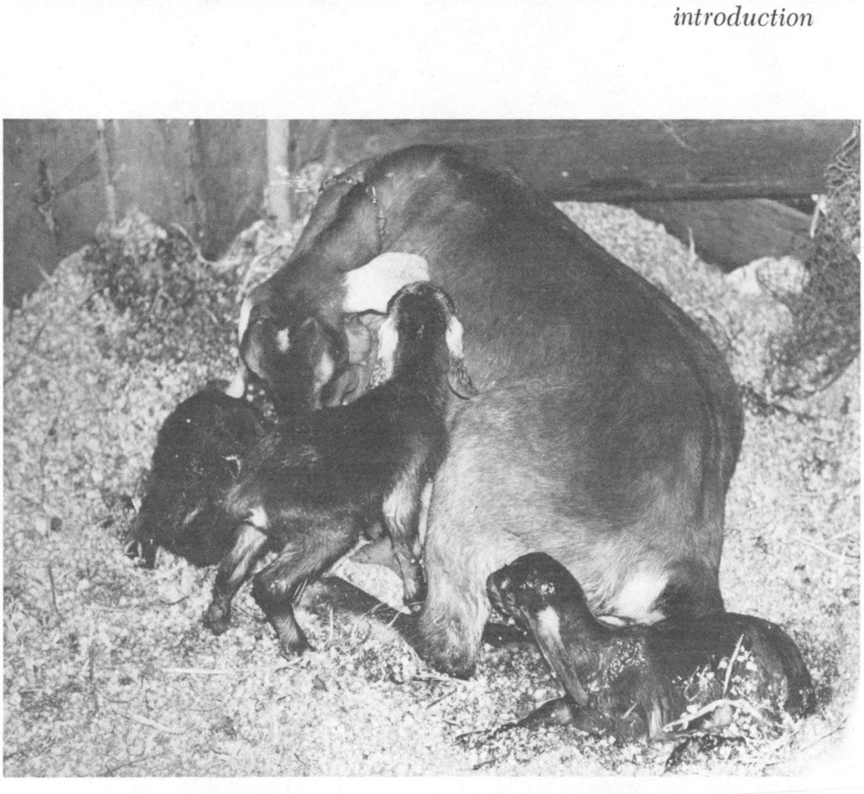

The homestead is a good place to learn about the miracle of birth and the interrelationship of life and death.

this is one of the attractions of raising small stock. The challenges—and the resultant rewards—are tremendous.

There are no experts because what works for you might not work for me. Talk to 12 experienced people and you'll get a dozen different answers. To progress, take those answers as a base, experiment with them, and come up with a thirteenth. Each animal is an individual. Each breeder is an individual. When you add to that different climates, different conditions within climactic areas, different feeds, different housing, even different sights and sounds and smells . . . well, there are just too many variables for any pat answers.

Of course you have to start somewhere, and there are basic rules and methods that are generally agreed upon. But please consider this guide as just that: a guide. A starting point. Follow the basic rules while you get to know your animals and their needs. Seek out help and advice wherever you can find it, and evaluate it in the light of your own growing experience. Beware of the "expert" with 20 years' experience who really has only one year's experience twenty times, and avoid becoming one of those yourself by developing your own methods and ideas, and by keeping an open mind. Similarly that means not telling a more experienced breeder he's wrong!

Above all, enjoy yourself and your livestock. Respect your animals. A good stockman is a gentle man in the way he handles his wards, in his consideration for their health and comfort.

Approached from this viewpoint especially, raising small stock is something you can enjoy all your life, and more with each passing year.

When I was in grade school, some of my best friends were pigeons. I still raise them, I enjoy their beauty, their personalities, their grace of flight. And one day I hope to be like the old man who sold me my first good birds. My life's work done, sitting in the shade of a vine-covered white loft, still thrilling at my birds—a little bit of me—winging their way to the clouds.

Jerome Belanger

SOME THINGS TO CONSIDER BEFORE YOU BEGIN

Of all the aspects of the current renewed interest in rural living, nothing holds more enchantment than livestock. The new ex–urbanite dreams of the doe-eyed, long-lashed Jersey cow and the bucketsful of rich milk she'll provide. He exults in waking as the dawn breaks to hear the crowing of his own rooster, and getting up to a breakfast of organic eggs, gathered clean and warm from his own nests. In his mind's eye, he sees hens scratching in the dust of the dooryard, hears the contented grunt of his fat pig lazing in the sun, is entertained by the antics of the young goats, and envisions fat sheep in the meadow and hutches full of plump young rabbits.

Livestock production, however, is not an enterprise to be taken lightly, to be approached without a good deal of consideration and study. No one will deny that rural living isn't for everybody. Similarly, livestock raising doesn't fit into the plans of every rural dweller. There are many points to consider before deciding whether livestock raising is for you.

The advantages of raising small animals are many, and most of them are apparent to anyone interested in the field. Economy is likely to be a prime consideration. Rabbit meat that sells for well over a dollar a pound in the market can be produced at home for less than one-fifth that price. Goat milk purchased for $1 a quart and more (when it's available at all) costs many homesteaders in the neighborhood of 10¢ a quart. Egg prices fluctuate wildly, but when they're 60 or 80¢ or $1.00 a dozen in the stores, it's nice to have a home flock that produces even better quality eggs for less than half that price.

The average family spends most of its grocery budget on meat and dairy items, so a home livestock enterprise should theoretically yield an even larger return on labor and capital than the home garden.

Naturally, if you are keeping your own animals, you have complete control over the quality of the meat, milk, and eggs you get from them. This control certainly gives you an advantage over those people who have to rely on food stores for their daily fare.

Organic milk, while difficult, if not impossible, to buy, can be produced at home by giving your animals only organically grown feed without regular feedings of medication. Raw milk cannot be sold in many states, but no matter where you live, if you're getting your milk from your own animals, you have the choice of pasteurizing it or not. Fertile eggs are expensive to buy, but there's nothing stopping you from keeping a rooster with the hens to fertilize your own eggs.

What's more, the flavor of home-grown pork can't begin to compare with anything bought in a supermarket, and the same holds true for chicken, eggs, and almost any other product.

Most small stock can be raised on a part-time basis; you don't have to be a farmer. With a job in town, caring for a few rabbits or goats can provide much needed relaxation and a change of pace, so the labor isn't really "labor" at all! Many small stock raisers would really rather spend time with their animals than on recreation that costs them money, so they in effect realize a double savings.

For the true breeder, there's a special thrill in developing *good* stock. Many a small, part-time farmer has turned fancier to show the results of his work and experience. There are local clubs and national associations that sponsor shows for rabbits, poultry, and goats. Besides meeting new friends with similar interests, showing is an excellent way to learn about a specific species.

In the final analysis though, all this is rationalization. No one raises small stock successfully unless they really love the animals they work with, and if they have that quality, none of the disadvantages of the project are likely to deter them.

And there are disadvantages! They vary with species and with individual situations, but there are undoubtedly days on any homestead when the caretaker wonders what in the dickens ever made him think this sort of thing was fun!

The most apparent disadvantage is in the responsibility involved, especially when keeping dairy animals. They must be milked every 12 hours, and unless a willing and capable neighbor can be enlisted, this means not only no vacations, but not even free weekends. In our footloose society, this alone is enough to discourage many people from getting started. (On the other hand, many goat keepers have told me, "When you have goats to play with, why would you want to go someplace else?")

All animals need regular care, of course, and this is something that should be seriously considered before deciding to commit yourself to them. It may be fun to carry water to the rabbits on a lazy summer Saturday afternoon when there's nothing else to do, but the rabbits will be just as thirsty when it's 20 below, one of your children just sprained an ankle ice skating, and you have three other couples coming in for dinner!

Animals are great for teaching children responsibility, as well as many other facts of life: reproduction, nutrition, genetics, economics, and even more. But the livestock project that starts out depending on child labor is on a shaky foundation. With some notable exceptions, children tend to lose interest, or at least find "more important" things to do occasionally. Mom or Dad must be prepared to handle this situation.

In the case of meat animals, you must consider the fact that your household will undoubtedly become attached to them. Will you be able to overcome this by the time the meat reaches the table? This is something most serious homesteaders grow used to without a great deal of trouble, even if the first few meals are somewhat unpleasant, but it's something to consider before you start.

In the same vein, will you be able to handle the admittedly unpleasant chore of butchering? Hiring outside help will wipe out much of the economic benefit of your project (as we'll seen when we discuss individual species), and doing it yourself takes time, a certain amount of skill, and the right frame of mind. These are decided disadvantages for many.

There are more subtle disadvantages, many of them dependent upon individual situations. Do you have enough land for the animals you're interested in? Can you grow feed, or will you have to buy it? Do you have suitable buildings, or will there be a construction project (and lumber bills) before the livestock project can take off? Negative answers to these questions can send a borderline homestead over the brink.

And there are still more subtle disadvantages once the farm is a going operation. There are goats that hardly milk, or dont' milk at all, and that inexpensive milk you dreamed of becomes very costly: goats eat whether they give milk or not. The same is true of chickens that don't lay, rabbits that don't reproduce, or sheep whose lambs die.

Obviously, there are pros and cons to any human endeavor. The intelligent person puts all the facts on a balance—usually adds at least a smidgen of prejudice one way or the other—and makes a decision.

* * * * * *

Once the basic decision is made to begin, there are still other general facts which apply to any class of livestock.

Any successful breeder has built on a foundation of good stock and has gone on from there to improve it. The attitude that a rabbit is a rabbit (or a goat is a goat or . . . name your main interest) can only result in disappointment. Cash "saved" by purchasing inferior animals is soon spent on inferior production. We aren't speaking of pets, but producing livestock, and no matter how "cute" or friendly or available, they won't do the job unless they're genetically capable. Upkeep is pretty much the same for good animals or poor animals. Original cost is soon overshadowed by upkeep. Production is the key! It means the difference between cheap food and food that you could buy at any store for much less.

A second axiom that applies to all animals is this: Poor management can ruin a good animal, but good management can't improve a poor animal. Starting out with good stock is important, but it isn't enough, because without good management that good stock can be run down into poor stock.

Management details differ for different animals, (and we'll cover these details later), but there are some basic principles that apply to all animals. Attention to detail is most important. The person who simply shoves the feed at his wards and thinks the chores are done is not a good manager. The person who breeds Susie the goat to Billie the buck just because he's handy, or Flopsy the rabbit to Mopsy for the same reason, isn't a breeder. He or she is merely a keeper and it won't be long before they aren't keeping much of anything worthwhile.

Attention to detail involves getting to know your animals. Is Susie eating less today than yesterday, and if so, why? Is her coat losing its sheen, and could it be because of the new mineral we're trying? Good managers observe how their stock eats, how they play, how they act toward other animals and their keeper. They observe the gleam in an eye, and even the droppings.

Detail also involves sanitation, one of the most important aspects of an organic farm. Most of the medications and chemicals employed in modern farming are used to combat unnatural living conditions, which generally means crowding and the resultant problems of sanitation. Animals confined by man depend on man to keep their surroundings pleasant. Many aspects of sanitation involve proper planning of housing, which we'll discuss in chapters pertaining to specific species. Good sanitation is one of the prime considerations of the home livestock unit.

One consideration often overlooked by the small farmer

is breed improvement. This is one of the values of shows. Many farm-type breeders simply have no interest in showing their animals, but this shouldn't be an excuse for not following the standards set up by breeders with years of experience.

It's true that many of the show qualifications for specific animals have no relationship to their utilitarian value. Horns are a disqualification on dairy goats, but you don't milk the horns. White (or colored, depending on the rabbit) toenails are disqualifications on some rabbits, which seems a bit silly when all you want is some good meat.

Yet, most of the points in the standards are there for a purpose, and it's pretty hard to break the rules intelligently if you don't know the rules in the first place. One of the most important tools in any breeder's box of tricks is the standard of perfection for his animals. That standard should be inscribed in his memory; everything he does should be aimed at developing that elusive "perfect specimen" described in the standard.

Some of the reasons are obvious to the experienced breeder. To a beginner, one rabbit looks pretty much like any other. The pro knows that the blocky shape of the New Zealand White rabbit is responsible for putting meat where it counts, that a snakey specimen dresses out with more waste and less desirable meat. The wide muzzle on a good goat, as called for in the standard, means she is a better eater; a good barrel means she has more capacity for feed, and feed produces milk; a well-shaped, capacious udder obviously means she at least has the tools to work with to produce the milk you're interested in. So even if you have no intentions at all of showing your animals, breeding to the standards means you stand a much better chance of getting the production you want.

There's another consideration here. Not only do many people who profess to have no interest in showing end up doing just that; sales of breeding stock can be a profitable sideline. For the most part, sales of such stock result from good show records, but in any event, the prospective buyer will be more impressed by animals that meet the standards. Many goat dairies or commercial rabbitries would be submarginal businesses if they relied on the sale of milk or meat alone: most of the gravy comes from the sale of breeding stock. One reason is that not every rabbit in a litter is a potential breeder; not every kid born in the goat barn is a keeper. Good animals are naturally worth more, and this additional income can be quite a boost to the small enterprise. The fact that showing is the best and fastest way to build a reputation should be of more than passing interest to the homesteader who thinks exhibiting stock has no place in his plans.

Management also involves detailed record keeping. At times, it seems like an unnecessary additional chore to weigh Susie's milk production and jot it down. The payoff comes in later decision–making. Even the best memory will be boggled when it comes to deciding which goat to keep on the basis of long-term production, or which rabbit should be culled because of repeated misses. Memory can play tricks, and you'll find yourself doubting your records much as the lost hunter doubts his compass.

Then too, records can be very impressive when it comes to stock sales. Detailed production accounts mark you as a conscientious breeder who knows where it's at and who's concerned with breed improvement, and as a good person to do business with.

If you are beginning to suspect that there is quite a bit more to successful small stock raising than feeding and hauling manure—congratulations. You stand a chance of becoming not a keeper, but a real breeder.

Yes, it takes work. It takes skill, acquired only by constant learning and through experience. But if the price is high, so are the rewards. The good breeder not only reduces his family's food bill considerably (and possibly even supplements it) but he earns satisfaction that can be found in very few other endeavors in today's plastic world.

RABBITS

WHY RABBITS?

As homestead livestock, the domestic rabbit has no equal.

The main goal of a home meat production unit is providing tasty, nutritious, chemically free food at a minimum of cost. In addition, for most people, it's necessary that the enterprise doesn't require a great deal of capital outlay, that it can be run without a great deal of experience, and in some cases, without a great deal of room. Would I sound prejudiced if I said the rabbit scores higher than any other domestic animal on all these points?

Many prospective homesteaders overlook rabbits as a source of meat, simply because Americans aren't accustomed to eating much rabbit. Beef—at least until lately—has been plentiful and relatively inexpensive.

But in many other countries where vast ranges for beef cattle aren't available, rabbit has been a staple for centuries. Frenchmen, for example, consume 13.5 pounds of rabbit per person per year. France produces 60 million pounds of rabbit meat annually, and Italy produces 115 million pounds.

In the United States, on the other hand, the per capita consumption is a mere two ounces.

No homestead food is worth producing if it isn't good tasting. I could no more describe the taste of rabbit than I could a prime steak; you'll just have to try it for yourself. The flavor is somewhat like chicken, but more delicate and subtle. Unlike the wild rabbit, domestic rabbit meat is all white. It's fine-grained, and when it's from a young pen-raised animal, exceedingly tender.

Just as important to the organic farmer, rabbit meat is nutritious. Although there is some disagreement with United States Department of Agriculture figures arrived at some years ago that placed rabbit higher in protein and food value than chicken, pork, or beef, rabbit still ranks closer in nutritional value to the red meats than to chicken. In infancy, rabbits are nursed on the richest milk produced by any animal: more than 15 percent protein, compared with cow milk which generally is about 3.5 percent. A young rabbit doubles its weight six days after birth, a calf doubles its weight only after 47 days.

Being lower in fat and higher in minerals than any other commonly used meat and being easily digested, rabbit is often prescribed for people with stomach trouble.

Besides being delicious and nutritious, homestead food must be easily and economically produced. Rabbit, again, walks off with the top honors. If, as we just mentioned, a rabbit doubles its birth weight in six days compared with 47 days for a calf, and there are usually eight young rabbits to a litter compared to one calf, the meat resulting from one breeding doubles in less than ONE DAY in the rabbitry.

To look at it another, more meaningful way, an 11-pound rabbit that weans 30 four-pound fryers a year produces 120 pounds of meat a year, or over 1,000 percent of her live body weight. A 400-pound brood sow that produces two litters of eight a year, with pigs averaging 25 pounds each when weaned at eight weeks of age, produces 400 pounds of meat or 100 percent of her live weight. A 1,000-pound range cow producing a 400-pound weaned calf gives a return of 40 percent.

Those 30 rabbit fryers, incidentally, are a conservative estimate and can easily be doubled when necessary or desirable. Moreover, while suckling pig and milk-fed veal are much more expensive (and presumably more valuable) than the meat from weaned animals, rabbits are commonly slaughtered without weaning. Not only do you get the higher quality found in suckling pig or veal, but it's actually more economical to butcher young rabbits—you get quality, and economy too.

There are other factors besides rate of gain that affect the price of meat, of course. Studies conducted by one feed company showed that it took 3.4 pounds of feed (including feed for the doe) to produce one pound of meat. If feed costs 5¢ a pound, the meat cost is 17¢ a pound. Most beef farmers figure their costs at twice this.

Then consider investment. While calf prices vary widely and have been fluctuating tremendously even within specific areas lately, a calf will cost at least $100. A just-weaned rabbit of good commercial quality can be had for $4-5, and while the cost of equipment will probably triple that figure, the total cash outlay is much less than for other livestock.

Naturally, rabbits require less space than other livestock, and while labor requirements are high, a few hutches can be handled in a few minutes of leisure time. Even children or elderly people can have fun doing the rabbit chores.

Perhaps even more significant in today's crowded world, rabbits can be raised in many places where any other livestock would be taboo. Properly housed, there should be no objectionable odor. They are noiseless. In most places rabbits are classed as pets, so even the homesteader who thought he was limited to a cat or a dog might be able to get into rabbit farming on a small scale.

And finally, we come to the "farm-retail spread." Ever wonder why a farmer gets, say, 35¢ a pound for beef on the hoof, and you pay three or four times that much? In 1971, a choice steer that brought the farmer $273.80, cost the consumer $427.98. This spread of $173.62 was up from $132.53 only five years earlier.

My family has raised pigs on our homestead that cost us around $40, and it cost us more than half of that to have them processed. We could have bought comparable hogs from local farmers at that time for $45. In other words, for six months of work, the risk of losing the animal through disease or accident, and investment in feed and equipment, we earned $5. For one afternoon of home butchering, we earned $21. And of course, going through a retail market would have added even more to the cost.

While home butchering hogs or cattle isn't for everybody, butchering rabbits is a cinch. My ten-year-old son is pretty good at it, and I've been in small producers' plants where the owners' teenaged boys did 100 an hour. What this means to the homesteader is that with a few minutes work, he can eliminate the middleman entirely by butchering rabbits, whereas he might not be able to with larger stock.

To look at the specifics on rabbits, I recently visited a processor who paid 27¢ a pound for live fryers. Pelts were selling for 2¢ a pound, but it cost 1¢ to ship them, so with the drying and other tasks, it wasn't worth it. He junked them. The dressout is about 50 percent, which automatically doubles the price of the meat. Processing costs came to 6¢

a pound, and the price to retailers was 60¢ a pound. The retail price in that area at that time was 89¢ a pound. (Frozen, shipped-in rabbit in Wisconsin commonly sells for $1.29 a pound.) Get the idea?

Rabbit raising as a commercial enterprise hasn't really gotten off the ground, anyplace. In France and Italy and other countries where production is far greater than in the United States, most rabbit meat is produced in the "homestead" fashion we're interested in. The operations are small, family units. In France, rabbit farmers can be found along the roads on summer evenings gathering grass for their animals. Table and garden scraps are common ration ingredients.

There are some large units in the United States and in England. The American rabbit industry is centered in Southern California, in the Ozarks, and there are sizeable markets in Southern Oregon and Florida. But new areas are emerging, and rabbit farming holds more promise today than at any period before.

So in addition to providing meat for the family table, there is always the possibility of expanding to a commercial farming unit. Most experts agree that it takes about 600 working does to provide a full-time job, or an adequate income.

SELECTING STOCK

To the uninitiated, a rabbit looks pretty much like any other rabbit. The dyed-in-the-wool fancier, however, through years of studying the Standard and rabbits themselves, can tell at a glance which rabbit is "better" than the next.

The rabbit farmer should fall somewhere between these

two extremes. Unlike the true fancier (unless he becomes one, of course, which is entirely permissible; homesteaders are human too, and deserve to have a little fun—and even the fancy rabbit raisers have plenty of culls for the dinner table!), the homesteader shows little concern over what color a rabbit's toenails are, the size or placement of a spot of color, or the shape of an ear. On the other hand, no serious rabbit farmer, no matter how small, would dare say one rabbit is as good as any other. That would be like a horse breeder entering a ragpicker's nag in the Kentucky Derby, just because it's a horse.

Perhaps the easiest way to explain this is simply to point out that rabbits are livestock. Since their discovery in Spain by the Phoenicians about 1100 B.C., they have undergone tremendous changes, some of which are very important to the homesteader.

This process of selection is as interesting as it is important. Consider for a moment that in the beginning there was a wild European rabbit, Oryctolagus Cuniculous. It was somewhat small, brownish (more correctly called agouti), with long ears. The Romans kept these animals in walled gardens to protect them from their natural enemies. The meat was said to improve a woman's beauty, and the embryos were considered great delicacies.

But also, given the rabbit's productivity and man's proclivity to experiment, entirely new breeds were developed from Oryctolagus Cuniculous. This didn't really get underway until the latter part of the Middle Ages, when French monks started keeping rabbits in protected surroundings and began breeding them selectively.

Natural mutations had presumably always occurred, but without selective breeding, the mutations were lost. With selective breeding, not only was the larger rabbit able to produce larger than normal young, to take just one trait, but the process could be carried on until it reached its natural limits, and rabbits of 20 pounds and more were being raised.

Today there are more than 75 varieties of rabbit descended from the wild European one. They bear no resemblance at all to the wild American rabbits. The domestic rabbit has all white meat which is delicately flavored, while the wild rabbits and hares (which are still something else) have dark meat which is usually "gamey" tasting.

Domestic rabbits come in mature weights of from two and one half pounds (the Netherland Dwarf) to 15 pounds and up (Flemish Giants). They come in practically every fur color imaginable, and some combinations that are pretty hard to imagine! (For example, the Harlequin, which in one color scheme has a head that is black on one side and orange on the other, a black ear on the orange side of the face and an orange ear on the black side, one front leg orange and the other black, and the hind legs just the reverse of the front. To add to the breeders' challenge, they must have hazel eyes, and white toenails are a disqualification!) Domestic rabbits are bred with long ears and relatively short ears, with arched bodies, compact bodies, and snakey bodies.

Ah, but you say, you're only interested in a few rabbits to eat. Isn't any rabbit edible, and if so, does it make any difference which breed you choose, just so it's big and meaty?

That's the whole point. Through genetics, animals can be molded to meet specific needs or desired goals. Just as the corn or tomatoes you plant in your garden bear no resemblance to the wild crops man first domesticated, and just as there are different breeds of plants to meet different needs—Roma tomatoes and the Beefsteak varieties, or field corn, pop corn, and sweet corn—the rabbits you choose to raise will bear no resemblance to the wild rabbits they descended from and certainly not to the wild American rabbit with which they have no connection.

So, if all you want is good meat for your own table, what breed of rabbit do you look for? Contrary to widespread popular opinion, the biggest isn't always the best. The 15- to 20-pound Flemish Giant is too often the first breed that attracts the attention of the potential rabbit farmer. Size alone is no criteria. The Flemish Giants are certainly edible, but their lack of fine bone and their heavy pelts as compared to the "commercial" breeds means you're putting feed into them that will not be converted to meat, but to waste. (The term "dressout percentage" applies to the percentage of edible meat to offal, which is the head, feet, and hide). Larger breeds take longer to mature, which means you feed prospective breeding stock longer before getting them into production. They also eat more. The Flemish aren't as productive as certain other breeds, again, because they simply haven't been bred for it.

We also hear a great deal about the Belgian Hare. (Actually, it is not a true hare at all, but a rabbit. Hares are larger than rabbits, their hind legs are longer, they do not have their young in underground burrows like true rabbits, and the newborn have a full coat of hair and open eyes when born, while rabbits are born blind and naked.) Interest in Belgian Hares dates back to the early 1900's, when some promoters with Belgian Hares to sell started an honest-to-goodness boom that's still remembered today. Fantastic prices were paid for Belgians (and even for some wild rabbits—to many people, a rabbit was a rabbit). Fortunes were made. But the poor people who bought their rabbits soon discovered that Belgians weren't bred for meat production;

they are strictly show animals. And when the bubble burst, those fortunes were lost even faster than they had been made.

So we still hear about Belgian Hares, and there are still plenty of con artists around trying to duplicate that Belgian Hare boom with their offers of unbelievable profits raising rabbits.

As a matter of fact, the type of rabbit we, as homesteaders, are looking for, wasn't developed until after the Belgian Hare episode. A few breeders who truly believed in the rabbit stuck with it after the crash, and more than that, they set to work to develop the kind of rabbit they thought was needed to produce meat efficiently. The Belgian Hare was part of the foundation stock, but what they came up with was the New Zealand, and later and more important, the New Zealand White.

The New Zealand is by far the most popular rabbit in America today. The American Federation of New Zealand Rabbit Breeders is far and away the largest specialty club in the country, and one-third of all rabbits registered with the American Rabbit Breeders Association are New Zealands. Commercial rabbitries raise either New Zealand Whites or Californians, with New Zealands far in the lead. The reasons for the New Zealand's popularity are many and varied, but they should prove to any doubter that a rabbit is not just a rabbit!

Being bred for production, commercial-type rabbits have obvious advantages for the homesteader. The fancier looks for one or two good babies in a litter; the meat farmer needs seven or eight uniform ones. The fancier doesn't want to push his stock—four litters a year is fine. The farmer wants five, or six, or even seven (and some people who are in the business for money push their herds even harder than that).

The farmer wants good dress-out. Belgian Hares can't be depended upon to deliver these objectives, nor can Flemish Giants, nor most of the other "fancy" breeds.

This means that the commercial rabbit has the stamina and vigor to raise large, healthy litters; the does are good milk producers; the animals have the correct body type to put the meat where it counts; and objectionable yellow fat has been bred out of the strain, as well as many common ailments and other faults that would cut into profits. In short, you have what is called the Hereford of the rabbit world: an animal bred for a specific purpose.

Incidentally, it's interesting to note that many people new to rabbit-raising are still interested in a market for furs, or they assume that rabbit furs are valuable. New Zealand Whites and Californians (which are white except for black foot and head markings which don't reach the pelt) were bred in white simply because of the fur market. Many processors who bought rabbits from producers paid a premium for white furs because they were worth more. They could be dyed any color, and most rabbit fur was used for trim. The fur market just about went out the window with the decline in popularity of fur and the advent of imitation furs for the small market that remained. The final blow was the drastic reduction in the manufacture of felt hats, which were made from rabbit skins. Today, most processors burn the pelts. In 1972, the price they got for a skin was less than the shipping cost.

Curiously enough, some processors still differentiate between white and colored furs. At least a few of them say it's because the white indicates a commercial breed, and they know it's likely to be a better rabbit that yields more and better meat than a colored one, which was not bred for meat production.

It would appear, then, that the homesteader who is seriously interested in raising rabbits and doing a good job of it should certainly investigate New Zealands. But there's more.

In any livestock, a superior animal of one breed is to be preferred to an inferior animal of another breed. In other words, even though New Zealands are supposedly bred for meat production and Flemish Giants (or any one of 40 to 50 other breeds) are not, it might well be that a very good Flemish Giant could outproduce a not-so-good New Zealand. Naturally this doesn't mean you look for a very good Flemish Giant, but for a good New Zealand. You see, even after you decide on a breed, a rabbit is still not a rabbit!

The problem for most beginners is knowing a good one from a bad one. Virtually everyone who isn't fortunate enough to live close to a topnotch breeder starts out with inferior rabbits, learns later what good rabbits are like, and moves up the ladder.

Here again, even though we may not have any interest in showing rabbits, it makes sense to take advantage of all the work fanciers have done in improving even the commercial breeds. Get a copy of the Standard of Perfection and study it for the breed you're interested in. Go to a rabbit show, and if the judge is any good, you'll learn something about why one rabbit is better than the next (not just for looks, but for meat production, too). And talk to breeders, which is easy to do at a show. Maybe there won't be any commercial growers there, but perhaps even a strict fancier can direct you to a reputable commercial operator.

In the final analysis, however, you're at the mercy of the seller. You want stock that is not only outwardly healthy and vigorous, which you should be able to determine for yourself by inspection, but also stock that has good size (not huge) litters, stock that raises a good percentage of those babies to slaughter age, and stock that produces good quality fryers. This rather simple request involves perhaps hundreds of genes. It involves everything from conception rates (missed litters and rebreeding take your time and feed money) to mothering ability and the genetic ability to produce firm meat instead of lanky bone or fat. The breeder knows what his stock can do, and you'll have to take his word for it.

In most cases, you won't expect your herd to do as well at your place as they did at their breeders. This is common among animals, and points out the importance of management techniques. Little things can make a big difference. But eventually, the stock will be accustomed to you and your way of doing things, and you'll be raising litters out of stock of your own breeding. This is when the selection of stock really begins.

SETTING UP SHOP

The wild rabbits of Europe from which our domestic rabbits have evolved lived in burrows, and the first attempts at domestication merely involved duplicating this natural habitat inside of walled gardens to protect the rabbits from natural enemies.

As in most other endeavors, it seems that significant change didn't come about until the twentieth century. The modern rabbitry is likely to have not only sanitary wire-floored cages for each mature animal, automatic watering (probably with a medication proportioner hooked in), and time-controlled, scientifically planned lighting, but even heating and air conditioning!

Obviously, the homesteader with three or four hutches will probably want something between these two extremes.

Every craftsman knows the value of good tools: they can make the difference between a good job and a poor one. The tools of rabbit husbandry are not only the stock, but also the equipment. The very best stock can be ruined by poor care, and sagging wire floors, or filthy solid floors, or water crocks that are so easily tipped the animal is usually thirsty, or improper nest boxes that contribute to the loss of young. All these and more amount to using poor tools, a handicap even skill and experience can't overcome. Good equipment is not a cost, but an investment, and the time and money it saves over the long run will make the higher initial cost seem insignificant.

It's generally a good idea to start at the top when considering equipment. Look at the best setup available, and if it isn't practical for your specific purposes or situation, then decide where to cut corners. In this regard, you could do no better than to visit as many local rabbitries as possible. (And I'll keep my fingers crossed that you'll be able to find a *good* one! Many so-called rabbitries are a disgrace to their owners and a black eye for the entire rabbit-raising fraternity.)

Naturally, there are local variations in the type of facilities needed. In the warmer sections of the country, even large rabbitries are commonly rows of hutches in the open or perhaps under some form of shade. Total environment facilities are on the increase even in those areas, however. Not only is heat a greater danger to rabbits than cold, but most of the larger farms are in mild climates, and the larger farms are the ones that have more to gain from ideal housing.

Air conditioning has not worked out too well, but in areas that get hot, roof sprinklers are often used to keep down temperatures inside the rabbit buildings. An even better cooling method is the pad and fan system. This is installed in one end of a long, totally enclosed building. Water is recirculated from a trough at floor level to the pad material at the ceiling. It's sprayed or dripped on the padding, and it trickles back down to the floor trough. Large fans blow through the wet padding material, and the evaporative effect of the moving air effectively cools the rabbitry.

A fan and pad system obviously isn't going to be economical for a couple of hutches, but many small raisers simply obtain the same effect by hanging a wet gunny sack on the windward side of their cages in really hot weather. This trick can save a doe that's due to kindle in hot weather, and naturally, her litter, too.

Cold weather is less of a problem, so long as the animals are kept dry and out of drafts. Large rabbit buildings in the North should be heated in the winter, but primarily to keep the automatic watering system from freezing and for the comfort of the raiser. Ventilation is very important in winter, too.

The type of facilities you decide to use then are dependant not only on your personal tastes and financial position, but also on the size of your herd and your location. There's a wide range of possibilities even within those limitations.

Picking the Building

Starting at the top, perhaps (I say "perhaps" because there's a lot of disagreement among experienced breeders on almost every phase of rabbit raising!) the best arrangement in any climate is a fully enclosed building in which all wire cages can be hung. The ideal, of course, would be a building specifically designed for rabbits, preferably in the shade.

A very nice rabbitry with wooden hutches in an indoor building. Outdoor hutches are different in several respects, of course. And outdoor hutches can make use of such labor-saving devices as the feed hoppers which can be filled without opening the cage door.

Without artificial heating, cooling, or lighting, it should have plenty of windows to let in sunlight and fresh air during the winter.

Two of the most common buildings used are old hen houses and garages. Earth floors are preferable because they remain drier and are easier to clean. Concrete can be gotten *really* clean, but the job will have to be done more often. Wood will absorb urine and spilled water and will rot, as well as cause odor problems.

This quonset-type cage is simple to make. A further modification is a semi-quonset, which has a square back, and thus leaves room above the nest box, where many does like to sit.

Cages

In converted hen houses and garages, cages constructed of one-inch, 12-gauge galvanized-after-welding mesh can be hung from the rafters or ceiling. The heavy gauge welded wire is a necessity. Rabbits chew on poultry mesh, and although they don't tear it apart, they might injure their teeth on it. The cages are suspended from the ceiling with 14-gauge wire. Don't support them on wooden posts or benches.

As was mentioned before, wood and urine are incompatible, and furthermore, such supports allow snakes and rats to get too close to the cages. Even if a rat can't get into a cage, it will chew off the feet and legs of babies through the wire.

The generally accepted rule of thumb for cage size is one square foot for each pound of rabbit. A nine-pound doe, then, would be comfortable in a cage 36 by 36 inches, with at least 18 inches of head room. If you have short arms it would be a good idea to have the cage only two and one-half feet wide, and longer, so you can reach to the back of the cage. If you have a choice, make the cage larger rather than smaller. Even with one square foot per pound of mother, with six or eight babies growing up, you'll soon have wall-to-wall rabbits. Each mature animal needs its own cage. More than one will fight, and two does together can induce false pregnancy in each other.

This set-up approaches the ideal for the homestead. But what if no outbuilding is available?

Outside Hutches

An outside hutch, or a series of them, is certainly acceptable. Rabbits have been raised this way for years and will probably continue to be raised this way. But some of the principles just mentioned should still be kept in mind.

For example, wood is probably the basic construction material for outside hutches. But remember that rabbits chew on wood that's not protected by wire, and wire over wood tends to trap droppings, making sanitation difficult or even impossible. And urine-soaked wood stinks. Keep this in mind as you build, and you'll save much grief later.

Wooden floored hutches are certainly acceptable. The main complaint with them is that they must be cleaned

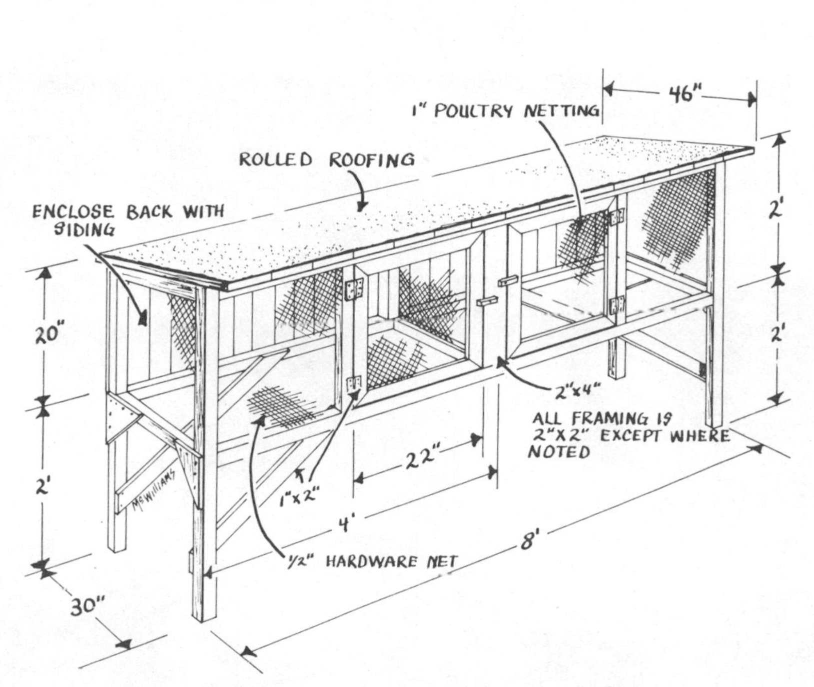

This is an ideal set-up for outside hutches. The roofing protects the animals from inclement weather, and the hardware netting on the front and sides permit good air circulation.

daily, a chore the commercial raiser, or even the busy homesteader, would rather avoid. Sawdust or shavings make good bedding, and you'll be amazed at what one rabbit contributes to the compost heap! But be prepared for that daily cleaning.

Raising Rabbits in Colonies

New rabbit raisers often ask about raising the animals in colonies, much like the walled gardens of the Romans we mentioned as being the first "kept" rabbits. It can be done.

This New Zealand White has a fine litter in an all-wire cage. The nest box has a removable top. It should be left in place when the babies are small so that the doe is less likely to jump on them and crush them.

In fact, *Countryside & Small Stock Journal* recently reported how rabbits could be raised this way with a minimum of labor and expense.

In one particular case, an area 16 by 16 feet was marked off and was dug out to a depth of two feet. At the corners, 12-foot posts were placed, two others were set at the middle of one side for a doorway, and center posts were set on the other three sides.

Bales of hay were stacked tightly in the depression, two bales thick. The baling twine was removed after the hay was in position. Poultry mesh was stapled to the posts, a hinged door set in place, and the entire thing was covered with black plastic for waterproofing. A large water trough was made from an eaves trough, and some oats, bone meal, and mineral was tossed on the hay.

One bred doe was placed in the enclosure. When her first litter was weaned, she was rebred and returned to the warren. After that, except for watering and feeding (grain and kitchen and garden waste), the rabbits were ignored for six months.

At that time, so the report went, the family of seven removed as many rabbit fryers as it wanted. They had rabbit nearly every day, and they had rabbits to sell! The rabbits tunneled into the hay and lived a fairly natural life.

For the homesteader who doesn't really care that much about rabbits, or who (mistakenly, in my opinion) isn't interested in stock improvement, this system could have some merit. There is no opportunity for selective breeding, which means the quality of the stock will be gradually lowered with a resulting loss of efficiency and meat quality. It is impossible to maintain any type of records necessary for such breed improvement anyway. So we pass this on, not as a recommendation, but as a point of interest and a possibility for those subsistence homesteaders who like to eat rabbit but don't care about raising them. So far as real rabbit breeders are concerned, this is just one step above going out in the woods and hunting rabbits.

Nest Boxes

In addition to the cage itself, you'll need a nest box. The nest should be removable, not built as part of the hutch. In

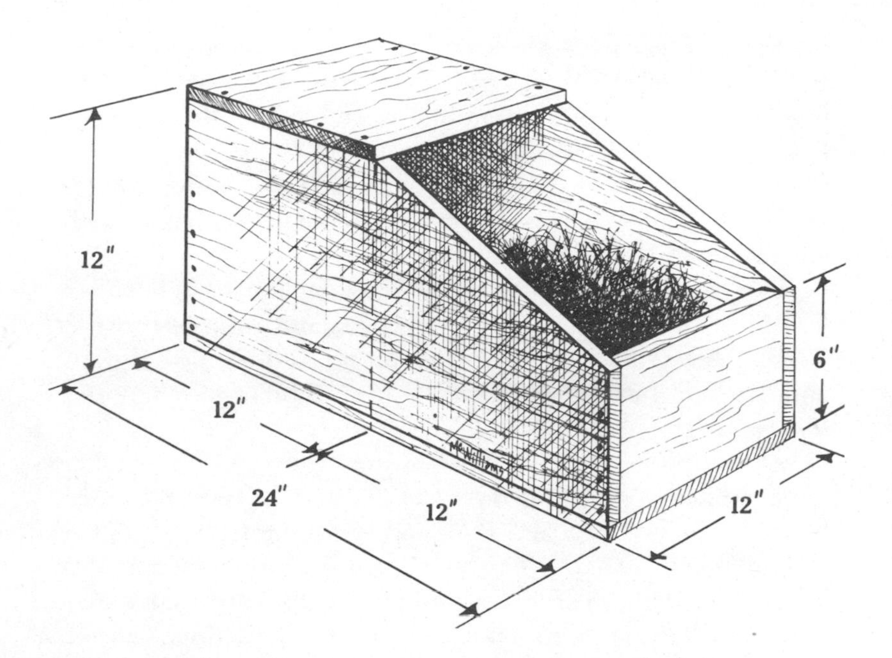

This simple nest box can be made from scrap lumber to fit right inside the cage.

fact, the buck doesn't need a nest at all, and the doe gets one only five days or so before she's due to kindle. She doesn't need it except to have babies in, and to have it in the cage after the babies are born just means added work for the keeper, because of the extra cleaning it requires.

The nest is generally about 12 by 24 inches by about 12 inches high for breeds such as New Zealands. They can be made of wood or sheet metal. In the summer some large rabbitries use wire nest boxes lined with paper. The paper is merely disposed of, and cleaning and sanitation labor is kept to a minimum.

The nest should have a top. Mother rabbit will enjoy

sitting on it, and it provides a good place to get away from it all (meaning eight or so playful, squirming little rabbits) every once in awhile. The top will also help keep the babies warm in cold weather.

There are two general types of nest boxes: one with the entrance hole cut at the top of one end, and the other with the entrance at one end of the top. In either case, the hole is about six inches from the floor to prevent the babies from being dragged out, or from getting out before they're smart enough to know how to get back in. Also, the mother is less likely to trample her young if she has to leap up and into the nest.

Metal nests can be purchased from any rabbit supply house whose address can be found in the rabbit or small stock magazines. Someone handy with sheet metal could build his own based on the general specifications for wooden ones, given here:

Cut three one-inch boards 12 by 24 inches. One is the floor, the other two are sides. Cut off one corner of the sides so the angle runs from about six inches from the floor back to about 12 inches from the front of the box. Then use a 12-by-six-inch piece for the front, a 12 by 12 inch piece for the back, and a 12-by-12-inch piece for the top. If you're handy with tools, fix it so the top and bottom can be removed to facilitate cleaning between litters. The edges can be protected with metal to prevent the rabbits from chewing on them, which will save you time in the carpentry shop building new nests.

The commercially available metal boxes have perforated hardboard floors which rest on flanges formed by the sides and ends being angled underneath. These floors are easily removed for cleaning.

Nests, especially wooden ones, are best cleaned by

scraping and scrubbing, then lightly burning with a torch. This destroys hair that soapy water and elbow grease can't remove and effectively disinfects the nest. An airing in the sun is a good idea then too.

Watering Equipment

Even more important than the nest are the feed and water utensils. Water—"the cheapest and most important feed"—should be available at all times, and it should always be clean and fresh, and the proper utensil is important if this goal is to be met.

During warm weather, a doe and her litter will drink as much as a gallon of water a day. A water dish, then, should hold at least one-half gallon. A coffee can or similar vessel can be used, and even though it's not ideal, it is inexpensive. Any sharp edges should be turned in and hammered smooth. Most rabbits will delight in picking up such water dishes and dumping them over as fast as you can fill them, so fasten them down securely!

A much better method of watering is to use half-gallon stoneware crocks made for the purpose. These have thick bottoms, which add enough weight so the rabbits can't toss them around, and concave bottoms and sloping sides, which allow the ice to rise as it forms in the winter, avoiding breakage due to freezing.

Still another type of watering system involves a metal trough, inserted through a 2-inch hole cut in the wire of the cage front. The shape of the trough protruding outside the wire allows it to be filled and emptied without opening the cage door. This is a real time saver if you have more than a few hutches.

Liver coccidiosis is almost impossible to control com-

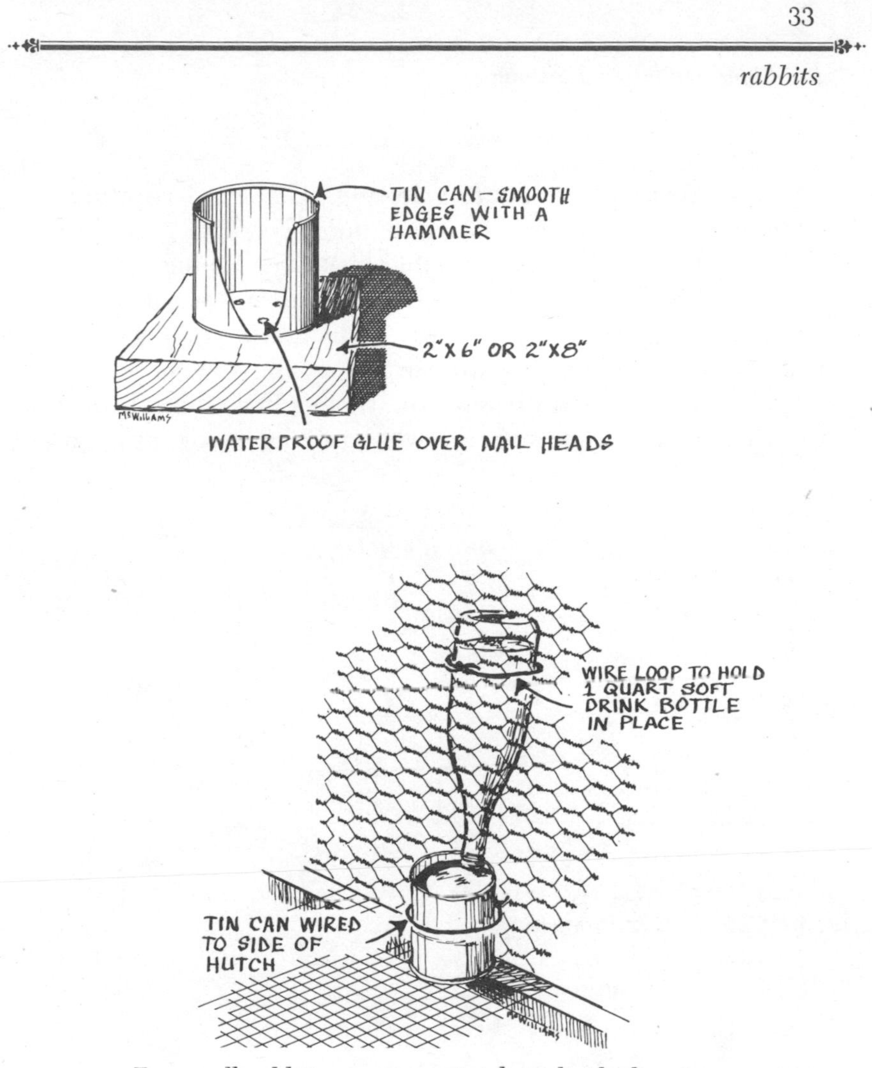

For small rabbit operations simple individual waterers can be made from tin cans and glass bottles.

pletely with any of these watering methods. Even with the best of management, dirt and droppings will get into the water to contaminate it. Furthermore, hand watering can be a tremendous chore: one breeder reported spending four hours a day—365 days a year—just *watering* her herd of 200.

The answer is automatic watering. There are cup-type waterers on the market, but these are much less desirable than the dew drops, a nozzle affair that even young rabbits learn to drink from with surprising speed and ease. Automatic systems involve a pressure reduction system with a float valve to reduce the pressure from the regular water source. With too high pressure the rabbits have a hard time drawing water. A pressure breaker can be made using a

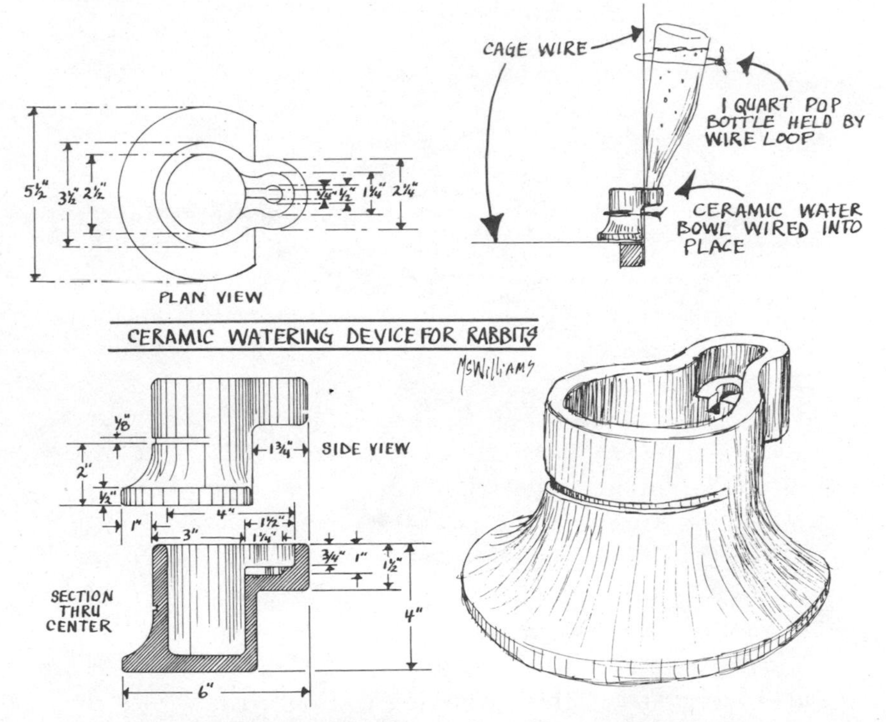

Ceramic watering devices, such as the one shown here, can be used for individual waterers. The ceramic bowl is too heavy for rabbits to dump over, there are no sharp edges to hurt them, and there are no metal parts to get rusty.

barrel with a float valve, but this will take some experimenting, because too little pressure will cause the dew drop valves to leak, and with too much pressure the rabbit can't trip the valve with its tongue.

From the pressure reduction tank, a half-inch pipe (and plastic pipe is much easier for the non-plumber rabbit raiser to work with) is run past each hutch. The supply pipe is outside the back of each hutch so any dripping will not wet the rabbits. About nine inches above floor level is right for medium-sized breeds such as New Zealands. This sounds high, but the young rabbits stretch easily to reach the water. With plastic pipe, the dew drop valves are merely screwed into the pipe at the desired locations.

Of course, there are drain valves and vents to eliminate air bubbles and a few other more or less technical considerations. Anybody serious about a system like this would certainly want to investigate more fully than the homesteader we're concerned with here, but, once again, the ingenius homesteader can take a leaf from the page of the successful large operator.

The full-scale automatic watering system isn't practical for a small set-up, but some of the benefits, particularly a constant supply of water and a supply that is impossible to contaminate, can be had by improvising on the larger system.

The dew drop valves themselves are inexpensive and they can be inserted into plastic jugs, such as bleach and other household items come in. A rack is made to hold the jugs upside down (you need air vents in the bottom of the jug, which becomes the top of the waterer). Watering becomes a once-a-day chore as long as it doesn't freeze, but even more important from the standpoint of good management, the rabbits never run dry, and their drinking water will always be fresh and clean.

The handyman might see in this an opportunity to go even a step closer to the fully automatic system, with a larger supply tank and the usual pipe running to each hutch. This is fine, and although the large tank can be filled with a hose,

Metal feeders are available in a number of sizes and styles. Some have mesh bottoms to permit the feed dust to shift through and therefore, not irritate the rabbits' delicate respiratory system. Note the space for the hutch record card, an important part of rabbit-keeping.

it shouldn't hold so much water that it becomes stagnant. Fresh water should be drawn each day, especially in very warm weather.

Feeders

As far as feed utensils go, coffee cans can also be used as feed dishes, with most of the same limitations they have as watering dishes. Again, crocks are handier.

The almost universally accepted feeder is a metal box which attaches to the outside of the pen. A trough portion goes through a hole cut in the wire. Some models have screened bottoms to permit "fines" (the dusty portions of feed) to sift through. The rabbits will not eat the dusty feed,

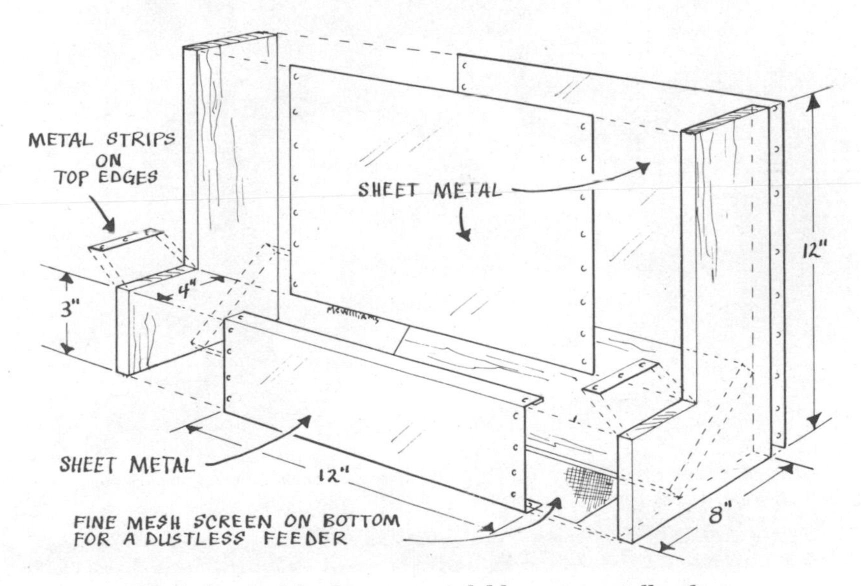

Although metal feeders are available commercially, they can be made at home from scrap lumber, sheet metal, and fine screen mesh.

and, if left in the case, it can cause respiratory problems. A rabbit breathes through its nose—it cannot breathe through its mouth—and its nose is very complex and sensitive.

Like the trough waterer, this feeder enables the keeper to go down each aisle and feed each animal without the bother of opening each cage door. For does with litters on full feed, these hoppers have the added advantage of holding more feed than crocks, with less waste. Baby rabbits, particularly, are prone to think a feed crock is a potty seat.

Even many simple rabbit raisers who feed complete pellets provide hay also. A simple hay manger can be seen behind this New Zealand doe.

Automatic feeding of rabbits has not been developed mainly because there doesn't seem to be any way to handle pellets without undue breakage and the attendant waste.

Hay Mangers

Hay mangers are not in common use in commercial rabbitries because the usual commercial ration is a complete pellet, which contains hay. Feeding anything other than pellets would be nearly impossible for the 600 or more working does needed to provide a full-time income. Fanciers, however, often claim that extra hay is an important

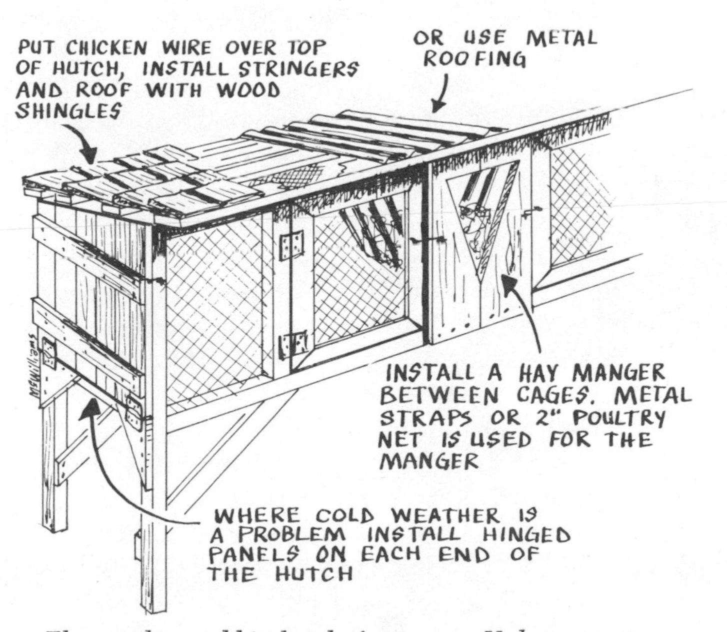

The outdoor rabbit hutch features a V hay manger which allows you to feed two cages at the same time.

part of their conditioning programs. Homesteaders will no doubt want to make use of hay, as we'll see in the chapter on feeding.

Hutches can be constructed with V-shaped mangers between each two units. The end is open, and the caretaker can go down an aisle and feed two hutches at a time simply by stuffing hay into these mangers. With all-wire cages, the manger can be built to fit inside, if the cage is large enough. It can also be fastened outside the cage, and the rabbits can pull hay through the wire.

Other Important Equipment

Although it may not seem as basic as a feeder or waterer, one of the most important items of equipment in the rabbitry is the hutch card holder. The hutch card is where you record the rabbit's life history, an absolute necessity when it comes to culling, intelligent breeding, and selection of future breeding stock. This card lists the sire and dam, date of birth and other pertinent data, date of each breeding, buck bred to, (if a doe) number born in each litter, number weaned in each litter, and weight of the litter at weaning and perhaps at other selected times. Commercial feeders have two flanges on the outside to accommodate these hutch cards, which can be hand printed or are available already printed from several of the larger feed companies.

Another must item for most small rabbitries is a metal garbage can in which to store feed. You don't want to encourage rats and mice in your rabbit barn, and you certainly don't want to feed them.

A scale is handy. A regular kitchen one will do, although it may take some patience to convince a young rabbit to stop wiggling long enough so you can take a reading. A hanging type is better. Suspend a basket from it and place the rabbit in the basket.

Another item you may want to include, at least when you get into operation and can see its value, is a small propane torch for cleaning nest boxes and cages. (That fur is terrible stuff to clean off in any other way, and manure tends to cling to it.) A measuring cup or small tin can that holds a known amount is useful for doling out the proper amount of feed, generally about five ounces a day for bucks and does without litters. Of course you'll need various cleaning tools, depending upon the type of hutches used, but a garden hoe with a sawed off handle is useful for a variety of jobs.

FEEDING

Feeding is easily the most important (and most expensive) part of livestock raising. As much as 75 percent or more of the total cost of raising meat involves feed. Not only is feeding expensive, but for the conscientious breeder, especially if he's interested in organic feeds, it can be quite complicated.

Some people no doubt will be surprised to learn that feeding rabbits is either expensive or complicated: they expect to toss them a few lettuce leaves or a carrot, and let it go at that. Farmers like that don't usually stay in the business very long.

A Balanced Diet

Rabbits are livestock and, as such, need a balanced diet with adequate vitamins, protein level, minerals . . . everything, in other words, the health-conscious organic gardener looks for in his own diet. Diet is especially important for nursing stock and rapidly growing youngsters.

Rabbits are vegetarians, but that doesn't mean they can thrive, or even survive, on salads. A doe and an eight-week-old litter would need about 50 pounds of cabbage a day, for example, to get the protein they need. That amount would obviously kill them, even if they could eat it. Fifty pounds of cabbage contains about six gallons of water!

And of course, this is just considering protein, but we know that the *type* of protein is important, too. While protein was formerly thought to be the basic building block, (*protos* means first, or basic) it has since been broken down into amino acids—some 23 of them. Some of these amino acids can be synthesized in the stomach. About half cannot be "manufactured" by the animal. These are called "essential amino acids" and must be present in the feed.

Obviously, the easiest way to make certain your animals have all the nutrition they need, at least according to the best knowledge of today's science, is to feed a commercial prepared ration. For rabbits, this is a complete, pelleted feed.

The major ingredient in rabbit feed should be legumous hay, generally alfalfa. The grains are generally selected on a least-cost basis, which is affected by location, time of year, and other factors affecting price. Minerals, supplements, preservatives (yes, animal feeds often contain preservatives too), and sometimes medications, are all mixed into the ground hay and grain.

But such ground feed is moistened with steam, forced through dies which shape it like spaghetti, and the "strings" are cut into three-eighth–inch lengths. The result is a hard, dry pellet which is palatable to the rabbit, and supposedly contains everything the animal needs to thrive.

For the commercial rabbitry, pellets are the only economical way to feed because of labor requirements. For the homesteader, especially the organic homesteader who wants the meat he eats to be the best possible, there are several alternatives.

Grains & Hay

For many small farmers, the biggest problem in feeding stock organically is obtaining organic grain. Not enough large acreage farmers are organic, and the small place too often can't support feed grains. You couldn't turn a combine around on the fields most homesteads require! Hand harvesting is possible, but impractical for most people, especially those part-time farmers who must spend their days at other jobs.

And then too, of course, many rabbit raisers are in such locations that there simply isn't room for growing crops. The answer to this problem is highly variable, but the point is, the main ingredients of rabbit feed are hay and grain. Don't plan on feeding rabbits from an ordinary vegetable garden.

Home-grown feeds obviously won't be pelleted. This means you'll need hay racks for each cage, in addition to grain feeders. Wooden hutches are often arranged so that one rack serves two cages, thus cutting the time involved in this chore in half.

The grains used could include oats, corn, wheat, milo or other grain sorghums, and barley. The selection is based on availability and the mix needed to meet nutritional requirements. (See appendix.) All cereal grains can be fed whole, but corn should be cracked to avoid a lot of waste, and oats and barley should be rolled. If ground feed is used (and there are some advantages to this) it should be moistened just before feeding to eliminate dust.

Supplements

Generally speaking, the greater the variety of feedstuffs included in the ration, the more complete it is likely to be. However, even a good selection of grains and hay, grown on well-fertilized soil, will not supply all the nutrient requirements of high-producing stock.

For example, protein, once again, is a major requirement. The recommended level ranges from 12 percent for maintenance (dry does and bucks) to 17 percent or more for pregnant does and does with litters. If corn contains about 9 percent protein and alfalfa hay about 15 percent and oats about 11 or 12 percent, a mixture of the three is going to require some additional protein to bring it up to the necessary level for good performance.

Soybeans are a good source of protein, and one easily grown on the homestead, but whole soybeans are not palatable to rabbits. They'll eat about one pound of soybeans to 10 pounds of other grains, but this will still require the addition of soybean or cottonseed meal.

Vitamins and minerals are generally not a problem in the formulation of rabbit feeds. Not much research has been done in this area, but we do know that vitamins A and D are especially important for rabbits. Rapidly growing plants, some of the root crops (carrots, for example) and good quality hay are sources of vitamin A. Small amounts of this vitamin are also found in yellow corn. Field-cured alfalfa is a good source of vitamin D. Most grains contain vitamin E in sufficient quantities, especially wheat. Rabbits apparently do not require vitamin C, according to the now defunct U.S. Rabbit Research Station.

The B vitamins are interesting. Ruminants (cows, sheep, and goats, which have more than one stomach) synthesize some of the B vitamins through the process of rumination. The rabbit is a single-stomach animal, but it has a form of pseudorumination, through coprophagy, or the consuming of some of its own feces.

Rabbits produce two kinds of feces: the familiar, large, "day" pellet, and a smaller, soft, mucous-covered "night" pellet. The latter is reingested directly from the anus. They may be found in the anterior portion of the stomach of healthy rabbits, intact. This ingestion is not indicative of a deficiency of any kind nor is it a form of perversion. It is a natural means for the rabbit to obtain the maximum quantity of nutrients from its feed. Apparently, the process permits the animal to form its own B vitamins.

Home-Grown Feeds

Even the small landholder who must buy hay and grain from a reliable organic farmer can grow some of his own rabbit feed. Greens may be fed in reasonable quantities: no more than the animal will clean up in 15 to 20 minutes. Such succulent feeds should be introduced gradually to avoid dysentery or bloat, which can be fatal, especially in young stock. Cabbage is the biggest danger here. Lettuce, chard, kale, and all the other familiar succulents can be fed in limited quantities. If you remember that the hay and grains have the major essential vitamins and minerals, many of which are lacking in the succulents, and that the rabbit may neglect these important foods in favor of greens just as a child neglects vegetables if given the opportunity to eat candy, you'll be all right. Many home-grown root crops are also good feed. Carrots, obviously, but also potatoes, mangel beets, Jerusalem artichokes, and others.

Apples and even pears can be fed to rabbits, and of course they love to chew on twigs or pieces of branches you prune from fruit trees. Alder, willow, and similar branches also make good chewing. They provide the rabbit with something to do, possibly give them some nutrients, and help them to keep their teeth worn down. Fruit tree leaves can also make a good "treat".

Comfrey is another crop the organic homesteader is likely to be interested in. It probably deserves a section of its own in a book like this because of the ease of cultivating it as much as for its nutritional properties. Sunflower seeds is also a good crop; it's a good source of protein. Both comfrey and sunflowers can be grown even on city lots, and small patches can be decorative, easily harvested, and extremely useful. Comfrey is highly praised by many as a tonic for rabbits (as well as for people and other livestock).

Feed Requirements

With these basic concepts firmly in mind, let's get down to specifics. The feed requirements of various classes of rabbits have been calculated by the U.S. Department of Agriculture as follows:

Dry does, herd bucks, and developing young	
Protein	*12-15%*
Fat	*2- 3.5%*
Fiber	*20-27%*
Nitrogen-free extract	*43-47%*
Ash or mineral	*5- 6.5%*

Pregnant does and does with litters

Protein	*16 -20%*
Fat	*3 - 5.5%*
Fiber	*14 -20%*
Nitrogen-free extract	*44 -50%*
Ash or mineral	*4.5- 6.5%*

(See Appendix for further explanation of these terms.)

Protein is the most important part of a ration, but also the most expensive. There is no danger of feeding too much protein: it will just cost more. Feeding too little, on the other hand, will set your stock back, especially youngsters. A rabbit doubles its weight in its first week of life and increases its birth weight 28 times by the day it's weaned. Since protein is the growth part of the ration, this rapid weight gain demands large amounts of it.

The organic farmer knows better than most that the quality of grain and forage depends on the quality of the soil. Therefore, any suggested ration is based on averages. Furthermore, such suggestions may not apply in certain areas, at certain times of the year, or in certain years because of the availability and expense of ingredients. But such suggestions are helpful as a foundation on which to build a ration based on your own needs and situation.

The following rations meet the requirements listed as needed by dry does, herd bucks, and developing young:

#1

Whole oats or wheat	*15 pounds*
Barley, milo or other grain sorghum	*15 pounds*
Alfalfa, clover, lespedeza, or pea hay	*69.5 pounds*
Salt	*0.5 pounds*

#2

Whole barley or oats	*35*	*pounds*
Alfalfa or clover hay	*64.5*	*pounds*
Salt	*0.5*	*pounds*

#3

Whole oats	*45*	*pounds*
Soybean, peanut, or linseed pellets, or peasize cake (38-43% protein)	*15*	*pounds*
Timothy, prairie or sudan hay	*39.5*	*pounds*
Salt	*0.5*	*pounds*

The higher protein rations for pregnant and nursing does should be based on the following:

#1

Whole oats or wheat	*15*	*pounds*
Whole barley, milo or other grain sorghum	*15*	*pounds*
Soybean or peanut meal pellets (38-43% protein)	*20*	*pounds*
Alfalfa, clover or pea hay	*49.5*	*pounds*
Salt	*0.5*	*pounds*

#2

Whole barley or oats	*35*	*pounds*
Soybean or peanut meal pellets or peasize cake (38-43% protein)	*15*	*pounds*
Alfalfa or clover hay	*49.5*	*pounds*
Salt	*0.5*	*pounds*

#3

Whole oats	*45*	*pounds*
Linseed pellets or pea-size cake (38-43% protein)	*25*	*pounds*
Timothy, prairie or sudan hay	*29.5*	*pounds*
Salt	*0.5*	*pounds*

A complete ration, from which pellets are made (but which homesteaders can feed ground and moistened), may contain these ingredients:

44% protein soybean meal	*18 pounds*
28% protein linseed meal	*4 pounds*
15% alfalfa meal	*40 pounds*
Wheat bran	*15 pounds*
Ground milo, barley, or corn	*18.5 pounds*
Ground oats	*4 pounds*
Salt	*0.5 pounds*

How Much to Feed

The amount to feed is variable, as different animals have different metabolisms. Here is where "the eye of the master" is important. If an animal leaves food from one feeding to another, cut back on the allowance. If it's always hungry, it might need more. However, overfeeding is as dangerous as underfeeding, as fat animals aren't producers. An overfat doe, for example, is likely to be a difficult breeder because fat builds up in the reproductive organs. And if she does get bred, she's more likely to have a difficult time kindling.

As an average, dry does in breeding conditions consume 3.8 percent of their live weight daily. In other words, a 10-pound doe eats 10 times 0.038 or 0.38 pound (six ounces) a day. This would amount to 2.5 ounces of grain and 3.5 ounces of hay. If green feed or root crops are fed, the amount should be limited to 1.6 ounces. Using this formula, the quantity fed can be adjusted for does of other weights.

Bucks and does under six months of age being developed for breeders will consume about 6.7 percent of their live weight daily. A rabbit that weighs four pounds at weaning will need about 4.2 ounces daily, but the quantity increases with the weight of the rabbit.

Feed enough to keep the animal in good condition, but don't "overcondition" by feeding too much. A bred doe and does with litters should be fed all they will eat without waste.

There's an old saying about the eye of the master fattening his stock—a saying which some people are trying to change to "the calculations of the chemist, the nutritionist, and the feed manufacturer's accountant fatten the stock." Most organic farmers will dispute this (and some honest scientists will back them up) on the grounds that we simply don't know everything there is to know about trace elements, interrelationships of elements, or even the needs of various animals. Not long ago, we thought proteins were the most basic feed elements. Now we have amino acids (which we don't thoroughly understand). Tomorrow?

Every animal lover knows that each animal is an individual, with particular needs and individual tastes. The commercial farmer can seldom afford to cater to the whims or needs of individuals, but the homesteader can, and will be rewarded for the extra attention. The good livestock breeder doesn't merely shove the feed at his wards and call the job done. Carefully observing eating habits, astutely keeping records of performance, and a certain amount of intuition make the difference between success and failure.

BREEDING

The beginning rabbit breeder all too often soon learns that the myth of rabbits' reproductive capabilities is just that; a myth. Quite often the problems begin with the actual breeding.

In the first place, a surprising number of people have trouble getting their rabbits bred. Our first suggestion is to tell them to make *sure* they have a buck and a doe! It's quite difficult to sex rabbits at first, and an amazing number of new breeders who have two bucks or two does wonder where all the babies are.

Sexing

On adult males, the testicles are quite prominent. If you get younger stock it will pay to buy them from a reputable breeder who will help you out and who will give you some information, too. In general, the method of sexing young animals is to hold one on its back supported by your left hand, with the tips of the fingers near the tail. Hold the tail with the index finger of the left hand to keep it out of the way.

Then, with the index finger and thumb of the right hand and the thumb of the left hand, press down firmly but very gently on the sexual organ. This will expose the membrane in the young rabbit. In the buck the organ will be somewhat round and will usually (but not always) protrude more than the female. The female sexual organ will form a slit. As the rabbits become older, the job is easier. The female has a more pronounced slit. The buck's male organ will protrude plainly when the genitals are pressed, and the testes will show quite plainly.

Mating

You'll need one buck for every ten does. In normal use the buck should not service more than two or three does a week.

The doe is taken to the buck's hutch for mating. If the buck is taken to the doe, she is more prone to fight to defend her territory. In most cases, mating takes only a few minutes, after which the doe is returned to her own cage. Don't leave them together unattended because they'll fight. Some breeders report a better conception rate if the doe is returned to the buck a few hours later.

There is no "heat period" as such in rabbits. The best research available indicates there is an estrus cycle of about 15 to 16 days, but in practice, because egg cells are developing and disintegrating in overlapping periods, the breeder need pay little attention to heat in rabbits. They can be bred practically anytime, providing, of course, the animals are in good condition. They should not be fat, or underfed, or in moult, diseased, or in any other condition which saps their vitality and will either reduce the probabilities of conception or will put undue strain on the animals.

Pregnancy and Kindling

The gestation period is 31 days, with some variation. During this period, the feed ration is changed to the higher-protein ration needed by pregnant does, and the amount of feed increased to what she will readily clean up.

About five days before kindling, the doe is given her nest box and a supply of straw, serval, old hay (not moldy), or any other handy nesting material. She will arrange the nest to her satisfaction.

The young are born with their eyes closed, and hairless. Most breeders like to check the nest to make sure the little tummies are full within a few hours of birth, to remove any dead, and to log in the number of young. The doe should be checked too, for general condition, caked breast (especially if the young have not been fed), and similar problems. Some does will reject their young if they are handled too soon after kindling, but most small breeders, who spend a lot of time with their does and handle them frequently, seldom have much trouble with this. If it becomes a problem, wait 24 hours before checking the nest, and/or dab a little Vaseline on the doe's nose so she can't smell you on her babies. Stroking her gently and speaking softly to her before checking the nest box will also help calm and reassure her.

The higher protein feed is continued at this point. Don't feed so much that it lays there and gets soiled. Feed just enough so that it's cleaned up from one feeding to the next.

The young rabbits open their eyes at about 10 days of age, and will soon be scampering about the hutch. At eight weeks of age, they should weigh four pounds or more, and are ready for slaughter. The doe can be rebred, and the entire cycle starts anew.

Rebreeding

A doe should not be left barren for a long period of time or she'll become permanently sterile. Breeding when the young are weaned will mean she'll produce four litters a year. If she averages seven per litter, or 28 a year, and you have three does, you'll have rabbit more than once a week which is just about right for the average homestead which also has other livestock.

If you like rabbit, or have a market for extra fryers, you might want to consider a faster rebreeding schedule. Some commercial breeders rebreed 21 days after kindling, some 14 days after, and some put even greater demands on their stock. The important thing here is determining the stamina of your animals. Some strains have been selected for such production, while others will soon burn out at a stepped-up pace. Go slowly at first and learn from experience just how good your rabbits are.

Feeding the Young

Many people ask about "fattening" young rabbits. If your stock is any good, this shouldn't be necessary, or even desirable. There is sure to be a temporary setback if the animals are weaned, and they require extra housing, care, etc. In most cases it just isn't worth it when they can be butchered at eight weeks of age. A four or four and one-half–pound rabbit looks small compared with a nine to 10-pound doe, but that's the best time to butcher them. The added weight gain after that will not compensate you for the feed used.

Creep feeding—providing a special ration the mother can't reach for young rabbits—is gaining in popularity among some raisers. It's been proven with sheep, hogs, and other livestock, but is comparatively new for rabbit raisers. And many who have tried it report extra investment and labor, with no extra returns in meat.

Keeping Records

Records have obvious importance in breeding. In the first place, you'll want to know the date of breeding so you know when to expect the young. You'll want to have the identity of both the dam and sire on record. The number of young kindled, the number weaned, and the weight at weaning are all valuable tools that will help make management decisions regarding culling, saving breeding stock, and future matings. Some bucks may work well with certain does and not with others. Attention to all these details are what makes the difference between success and failure in rabbit farming.

BUTCHERING

The purpose of rabbits on the homestead is for providing meat. While there is nothing cuter than a cage full of fuzzy small bunnies, you can't let yourself become attached to them! There are tears at many a large livestock show as a 4-Her's prize lamb or hog or steer is led away to be butchered, but this is an accepted part of farm life. Make pets of the working does, if you must, but decide from the outset that you're raising rabbits to be slaughtered.

Rabbits, in my opinion, are the easiest animals on the homestead to butcher. I've seen teenage boys process 100 an hour, but even at my speed (maybe 10 an hour on a good day) it's far faster and easier than plucking chickens.

The easiest way to stun a rabbit is with a good stick about the diameter of a stout broom handle and about 18 inches long. Restrain the animal with one hand, and hit it on the head, preferably just ahead of the ears. Some refuse to be restrained, in which case it may be necessary to hold the ears and then it's easier to strike it behind the ears. Both ways work.

A faster method of stunning rabbits, although it requires more experience, is to hold it by the hind legs in your left hand, and cup the chin in your right hand with the thumb behind the ears. A downward jerk with the right hand will dislocate the neck and stun the animal.

In both methods of stunning, you must remove the head to kill the rabbit and to assure thorough bleeding.

Once the head is removed, suspend the carcass from a gambrel hook (fashion one from a piece of heavy wire) through the tendons of either hind leg and remove the three free feet. A pruning shears works better than a knife for this operation.

Then cut the skin around the hock of the suspended leg and make a slit down to the groin. Slit the other hind leg in the same manner, and peel the skin off the legs. Cut it at the tail. Then simply peel the skin back until it can be pulled off the carcass inside out, like taking off a glove.

To eviscerate the rabbit, make an incision along the center of the stomach. The entrails are removed through this incision (take special care when cutting around the anus), and the liver, heart, and kidneys are saved. Carefully remove the gall bladder from the liver. Cut off the remaining hind leg, wash the carcass in cold water, and it's ready to be cut.

For appearance and ease in cooking and serving, rabbit is usually cut into seven pieces. With a good butcher knife make one cut just behind the shoulder, one just ahead of the rear legs, and one to divide the middle section. The leg portions are both split into two pieces, and the saddle portion of the middle section is cut lengthwise.

The meat should be cooled rapidly, but prolonged immersion in cold water will damage the appearance of the meat. It is better to chill it under refrigeration, if possible.

The hide is worthless on today's market, but it can be saved for home tanning. Fryer pelts are quite thin, but no doubt the ingenius homesteader can find many uses for them.

If a pelt is to be saved, wash it to remove blood and stains, and cut away any fat or tissue that may be adhering. Then stretch it on a frame made of an ironing board-shaped

Domestic rabbit meat is all white, fine grain, easily digested, highly nutritious . . . and delicious!

piece of wood, a wire frame made for the purpose, or easiest of all, a wire frame you make yourself from an old coat hanger. The hide is dried uncut with the flesh side out—just as it came from the rabbit. It should be hung in a cool, dry place—not in the sun. Don't salt it. Such uncut hides are called "cased."

RABBIT COOKERY

The real reward for the homesteader—the culmination of all his work and knowledge—comes in the kitchen. Those who have never eaten rabbit are in for a surprise, because it is not only versatile and highly nutritious, but delicious! The cut-up pieces look much like chicken (which helps squeamish members of the family), and the finished product often looks like chicken, too. In fact, rabbit can be cooked according to many chicken recipes.

But rabbit is *not* chicken. It has a much more delicate flavor, which many recipes tend to overpower. There's no accounting for taste, though, so just experiment and you'll surely come up with your own favorites.

Just remember that rabbit is a great delicacy in many restaurants, and that for many nationalities with a reputation for gourmet cooking, such as French and Italian, rabbit is a staple. Rabbit stew may be a peasant dish, but rabbit can also be a gourmet treat.

Here are some recipes to get you started.

FRIED RABBIT

1 cut-up rabbit fryer
1 small can mushrooms
3 small white onions
1 cup white wine
3 tablespoons butter
flour, salt, pepper

Dust rabbit meat lightly in flour, salt, and pepper.

Brown in butter on both sides. Add chopped onions and wine. Cover and simmer for 30 minutes. Add mushrooms and simmer for 10 minutes. Serve with gravy made from pan drippings.

CURRIED RABBIT
(Pat Katz)

1 young rabbit
1 tablespoon butter
1 tablespoon oil
1 chopped onion
1 large cubed apple
2 tablespoons raisins
1 clove crushed garlic
2 tablespoons coconut (dried unsweetened is all right)
¼ tablespoon chili powder
4 teaspoons curry powder
2 teaspoons thyme or savory
salt and pepper
1½ cups rabbit stock
1 large banana, sliced

Simmer the rabbit in water to cover until tender, about ¾ hour. Take the meat from the bones and cut into 1-inch cubes. (The bones can be returned to the stock to make it richer.)

Melt the butter and oil in a heavy pot with a tight lid. Add the apple, onion, raisins, garlic, and coconut and mix well. Add the chili, curry, herbs, salt, and pepper and cook for five minutes, stirring often.

Add the rabbit meat and stock. Cover and simmer gently for an hour. Add a little more stock if it becomes dry. Add the banana, stir, and cook for another 20 minutes. Serve with rice.

RABBIT CACCIATORE LEONE

(From Gene Leone, proprietor of Leone's Restaurant in New York City)

- *1 cut-up fryer rabbit*
- *3 tablespoons pure olive oil*
- *¼ pound bacon*
- *6 bay leaves*
- *1 cup finely chopped onion*
- *1 clove garlic*
- *¼ teaspoon freshly ground pepper*
- *4 cups canned stewed tomatoes*
- *¼ cup tomato paste*
- *¼ teaspoon crushed red pepper*
- *½ teaspoon oregano*
- *4 teaspoons butter*
- *½ cup finely chopped parsley*
- *½ teaspoon salt*
- *½ cup dry red wine*
- *olive oil*
- *flour*

Place the olive oil in a deep saucepan and put over medium heat. Chop the bacon finely and add it and the bay leaves to the hot oil. Add the chopped onion and cook until it is a little past golden brown. Chop the garlic and add it and the pepper to the saucepan. Cook until the mixture is well browned.

Add the stewed tomatoes, tomato paste, red pepper, oregano, butter, parsley, and salt. Bring the mixture to a boil, remove from heat, and let stand until the rabbit is added.

Flour the sections of rabbit and gently brown them in a separate frying pan in olive oil. When brown, add the red wine. Simmer for five minutes and then add rabbit and wine to the saucepan containing the cacciatore sauce. Simmer the rabbit in this sauce for 20 minutes, or until tender.

Leone's serves rabbit cacciatore with rice cooked in beef stock, and seasoned with saffron, freshly ground pepper, salt, and butter.

RABBIT POT PIE

1 3-pound rabbit
2 teaspoons salt
1 bay leaf
2 stalks celery
1 onion
½ cup parsley
4 tablespoons butter
4 tablespoons flour
3 cups rabbit stock
dash of Tabasco sauce

Wash and cut rabbit into pieces. Place in kettle, and add salt, bay leaf, and water to cover. Cover kettle and bring to boil. Boil until tender, then remove meat from bone and cut into large pieces.

Melt butter in skillet, add chopped onions, celery, and parsley, and cook until slightly transparent. Add flour slowly and mix well. Add the rabbit stock, and cook until thickened. Add salt and Tabasco sauce, mix well with rabbit meat, and pour into a baking dish. Cover with your favorite pastry, to which a few flakes of herbs may be added. Put slits in crust, and bake until crust is done.

LAPIN DIABLE A LA CREME

1 cut-up fryer rabbit
1 pint of cream
1 pint of white wine
10 large mushrooms
4 finely chopped shallots
1 teaspoon powdered mustard
2 pints beef stock
½ pound of butter
1 teaspoon arrowroot
1 black truffle
1 pound wild rice
croutons
salt and pepper

Melt ¼ pound of butter in a large pan until it turns to a golden brown. Sauté rabbit in the butter. When done, sprinkle the finely chopped shallots over the meat. Simmer for five minutes, then add the wine. Cover the pan and place it in the oven.

When the rabbit has cooked, remove the pieces and place them where they will keep warm. Leave the sauce in the pan.

Meanwhile, take one pint of cream and mix with the mustard and beef stock, making certain they are thoroughly blended. Add this mixture to the sauce, stirring constantly. Heat the sauce until it's reduced to the desired consistency.

Then add the arrowroot to the sauce, and stir until the sauce has thickened slightly. Pass the sauce through a piece of fine muslin and add to it the remaining ¼ pound of butter.

Cook the wild rice as you would a rice pilaff (basically, this is rice cooked in meat stock and seasoned with paprika, brown sugar, bay leaf, and onion, although there are many variations) and sauté the mushrooms in butter.

In the center of a large round silver platter place the rice pyramid style. Evenly distribute the pieces of rabbit around the rice. Then spoon your sauce over the rabbit. Place a mushroom head over each piece of rabbit and a slice of truffle over that. Decorate the border with croutons, and serve.

CHICKENS

THE MOST COMMON HOMESTEAD ANIMAL

The chicken is probably the most universally accepted livestock on the homestead, for psychological as well as practical reasons. Chickens require little space, a minimum of experience and knowledge, and they're readily available at low cost. People who think they could never eat rabbit or drink goat milk have no qualms about eating chicken, or certainly eggs.

Although a hen and her chicks scratching in the cottage dooryard is part of the dream of country living, the psychological factor goes even deeper today. The rise of "broiler factories" has been sudden enough that even many middle-aged people can remember what *good* chicken used to taste like, so they know how pallid modern chicken is by comparison. Eggs, too. In searching for a return to old-fashioned values and real food, the chicken just has to be an early entry in the homestead livestock lineup.

Archaeologists tell us chickens have been domesticated since prehistoric times. In fact, artificial incubation was used in the times of the Egyptian pharaohs. For those thousands of years, very few improvements or changes were made in the keeping of poultry. Even in this century, nearly every farm had a small flock of chickens whose care fell to the farm wife because they weren't "important" enough for the man to worry about. Those chickens got whatever grain the wife could borrow from the other animals, and in most cases, they were allowed to roam and complement their own diet with insects, worms, and vegetation.

The biggest and best young rooster was commonly the first killed, which meant one of the stragglers became the "herd sire". The effect of this kind of breeding program on the flock is readily apparent to us today: breed improvement was nil.

Both eggs and meat birds were commonly sold directly to the consumer, but meat was largely a by-product. As late as the 1930's, chickens were raised for eggs, and spent layers and young cockerels were used for meat.

The first American incubator was patented in 1844, but it was well into the 1900's before many changes started taking place. Hatcheries came into existence and became larger and fewer in number as time went on. In 1934 there were over 11,000 hatcheries in the United States. Thirty years later there were only 2,365—but they hatched six times as many eggs. In fact, about 600 hatcheries accounted for 71 percent of the nation's chicks.

Breed improvement started to take place, too. Through genetics, egg production increased from about 100 per year per hen 100 years ago, to over 220 today. (Scientific feeding and housing plays a role in this increase too, of course.) And special meat birds were bred, starting a completely new and separate industry.

With improvements in efficiency and lowering of price, chicken became a staple in America. In the 35 years from 1935 to 1969, broiler production increased 82 times, from 34 million to 2.8 billion! In 1910, 88 percent of American farms had chickens. The flocks averaged 50 laying hens. Today, of course, the situation is far different with flocks of 100,000 and more being common.

What has happened, then, is that through management techniques, today's chicken lays more eggs on less feed and with less labor. Meat birds have made similar advances.

And yet, the homesteader isn't sure all this is such a great blessing. A small home flock doesn't take much investment or time, but most important, birds raised the old-fashioned way just don't lay pale watery eggs like those in cages; and meat birds raised in batteries or under confinement just can't compare with the flavor of chickens raised in more natural environments.

Americans consume about 39 pounds of chicken and 313 eggs per capita per year. Both are important sources of nutrition. Eggs contain not only an abundance of protein, but an abundance of high-quality protein. They contain all the essential amino acids—the basic components of protein—needed to maintain life and promote growth. Eggs are a rich source of iron, phosphorus, and trace minerals, of vitamins A, E, K, and all the B vitamins. They are a source of vitamin D second only to the fish oils. As far as the meat of the chicken is concerned, it is higher in protein than beef and other red meats, and is much more easily raised and processed on the homestead.

GETTING STARTED

The poultry industry is comprised of breeders who sell eggs to hatcheries, who sell chicks to growers, who sell eggs or broilers to processors and distributors. But the entire job can be done on the homestead.

There are many ways to get started. Some homesteaders get surplus mature birds from neighbors, some buy spent layers from commercial egg producers, other buy started pullets, and still others prefer day-old chicks. Some even buy hatching eggs.

Before undertaking a poultry project and deciding upon a breed, the homesteader must decide just what he expects from his home flock. Raising chickens for eggs is different from raising them for meat, and raising chickens for meat is different from raising them to develop a breed for exhibition. The purpose of the enterprise must be decided before a breed can even be selected.

Most homesteaders, of course, want both meat and eggs. Many of them are disappointed to learn that you can't get both good egg production and meat production from the same birds. If you want top efficiency (a desirable goal on any homestead, and one that may be quite necessary if you can't produce your own feed), it might pay to consider getting two different breeds, one for eggs and one for meat.

The old-fashioned way is to have a dual-purpose breed that isn't really satisfactory for either purpose by today's standards. These breeds are not in demand by commercial farmers so they're a little harder to find. Many smaller hatcheries still provide them, however, and you can find a number of their advertisements in farm-oriented publications.

Another "problem" some homesteaders run into is that most of the modern strains of production chickens have the broodiness bred out of them. You won't get those fluffy little chicks scurrying about mother hen as you dreamed about. Broody chickens don't lay, so the mothering instinct is a liability to commercial egg producers.

Picking Your Birds

One of the most popular breeds of chickens for the homestead is undoubtedly the Bantam, not so much for any production attributes as for their charm. They are colorful, excellent mothers, and they lay amazingly large eggs, considering their size. Because of their small size they are somewhat of a bother to butcher, but their meat certainly is edible.

Bantams add a lot of country charm to the homestead. They keep garden pests in check, and their eggs are much larger than the size of the bird would indicate. They are also good brooders, and are even used to hatch goose eggs!

For utilitarian purposes, the larger breeds—even larger than commercial layers—are extremely popular among homesteaders. Of the more common ones, Rhode Island Reds and the White, Barred, and Plymouth Rocks do well both in the nests and in the frying pan.

Some more exotic breeds such as the Salmon Faverolle and Auracana have small but loyal followings. The Auracana has come into the limelight recently because of the colored eggs it lays. It's known as the "Easter egg chicken."

There are literally dozens of other breeds in this category. Looking over a catalog from a hatchery will undoubtedly fire up your desire for one or more unusual breeds.

Or, you might take advantage of some of that technology which has put a chicken in every pot in the United States. Follow the lead of the commercial growers and get some White Leghorns for eggs, and Cornish crosses for meat. There are many hybrids of both of these, but these two breeds form the basis for nearly all commercial birds in their respective specialties.

The Leghorns and Leghorn crosses are small birds, so they eat less, but they lay large eggs and in astounding numbers, so your cost per dozen eggs is much lower. They are not good meat birds, though, because of their small size.

The Cornish and Cornish crosses, on the other hand, have all the desirable qualities of good meat birds: wide breasts, yellow skin and white pin feathers, rapid growth and large size. They don't make good laying birds, though, because they obviously eat more than Leghorns and don't lay as well.

Some homesteaders have a veritable Noah's Ark in their henhouse: they have a few birds of many different breeds! A wide assortment of varieties can add a lot of interest to your chicken yard and might give you a good chance to decide which breed or breeds you prefer.

Once you have decided what breed or breeds you want, you have to decide what age the birds will be. You can buy hatching eggs, day-old chicks, started pullets, mature birds (from a neighbor who might be willing to part with some), or spent layers from a commercial flock. It's largely a matter of choice and availability.

Buying Day-Old Chicks

Buying day-old chicks is probably the most common way of getting started—and certainly the most fun! We've been raising chickens for a long time, but the day the post office calls to say our chicks have arrived is still one of the highlights of homesteading.

Everything is in readiness, since the hatchery alerts us to the impending arrival. We have an area with ½ square foot of space per bird deeply littered. (When we first started, this was a four-by-eight-foot piece of plywood, with jerry-rigged wooden sides about a foot high. We put this box in the print shop, since that was the only place we had that was warm enough. Some of the dust those chicks raised as they grew never *did* get out of the printing equipment! Aside from that, the location worked out fine.) We use a heat lamp for warmth, and check it carefully before the arrival of the chicks, so the temperature is 95 degrees at about two inches above the floor. The temperature is regulated, naturally, by raising or lowering the lamp.

Caring for the Chicks

For the first few days, a small fence is needed to keep the chicks near the heat source. We've constructed simple fences by taping cardboard together to form a circular hover. Metal would be more stable. This fence is circular to prevent the chicks from piling up in a corner out of fright from a sudden noise (like turning on a printing press) and smothering.

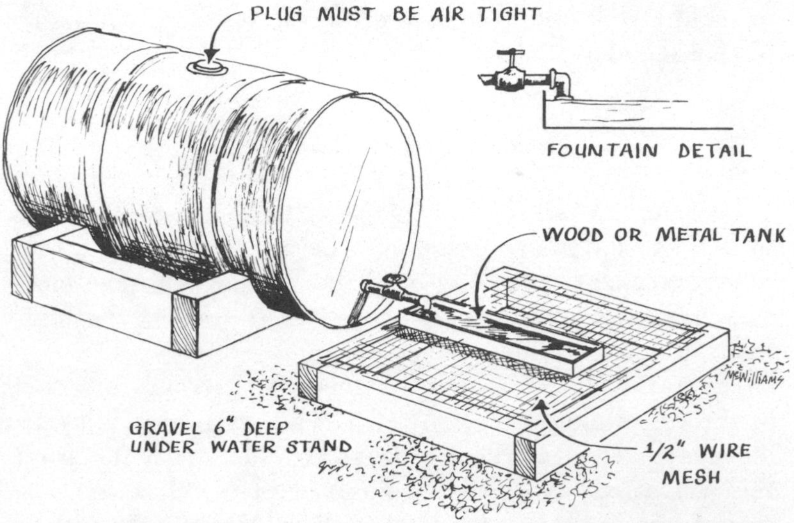

This watering set-up keeps the water in the trough clean and always available. Droppings go through the wire mesh, not in the water to contaminate it.

You need at least two one-gallon waterers for each 100 chicks. The small plastic chick waterers used with ordinary fruit jars are fine for a few birds.

You also need chick feeders and litter. The litter should not be dusty. Straw is not good litter. We always use ground corn cobs, but peanut hulls, peat, sugar cane fiber, and similar materials are good, depending on availability. A newspaper is spread on the litter, and some chick mash and scratch is spread on this paper. The chicks will eat litter otherwise, so this paper is used for several days.

It's a good idea to dip each new chick's beak in the water as it's removed from the box. In the first place, if they've been en route for a long time, they're probably dehydrated. And second, it teaches them early where to find

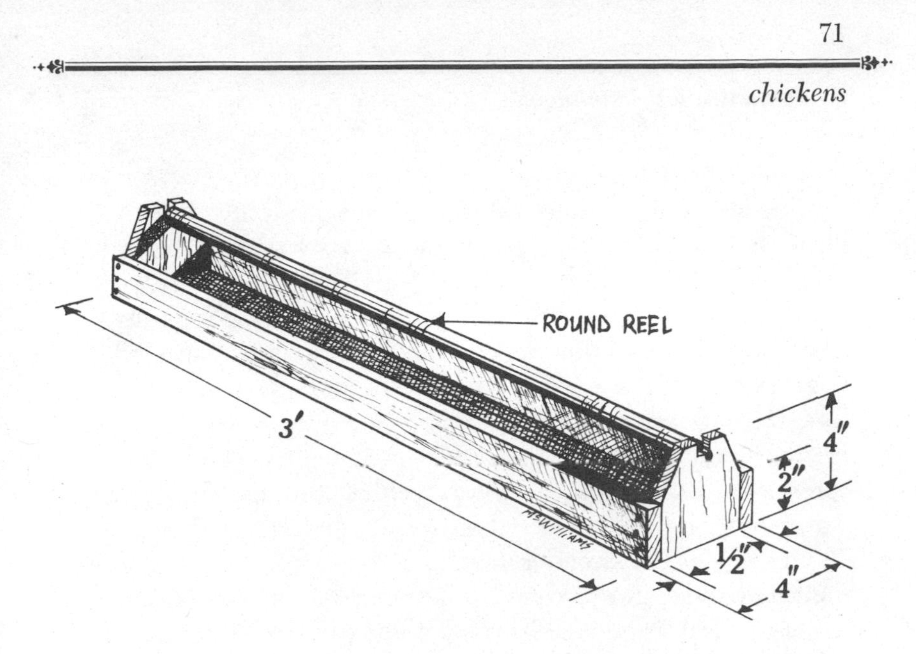

This feeder is large enough for about 50 chicks. The round reel in the center prevents chicks from stepping in their feed and contaminating it.

water. Make sure the chicks are in good condition. Most hatcheries include extras to make up for losses, but we've never had a dead bird arrive by mail.

Observe the chicks carefully and frequently after they arrive. See that they become accustomed to the heat. After they settle down, you can tell by watching them if they're too hot or too cold by the way they crowd under or avoid the heat lamp. Contented chicks are quiet, with contented chirps. When the chirping becomes shrill, you know something's wrong. If they all gather at one side of the hover, check for drafts.

The first 10 days are critical. Provide plenty of feeding space—at least one inch per chick to start with—and keep the feeders about three-fourths full. (If too full, a lot of feed is wasted.) Water is the most important food for any animal: keep it clean, and never let it run out.

Feeders and waterers should be washed daily. Turn the litter daily to keep it dry and clean. Use a night light to avoid crowding and smothering. This light should only be about 15 watts: too much light is one cause of cannibalism. The temperature of the brooder should be reduced about five degrees each week until it reaches the outside temperature.

After about a month, you'll need larger feeders and waterers, and at least ¾ square feet of space per chick. About three inches of feeding space per chick and one five-gallon water font per 100 chicks should be provided at this time. If you're using a brooder house, provide roosts, with four inches of space per bird and poles about six inches apart. If a clean yard is available, and the weather suitable, the chicks can be let outside.

This is just one method of brooding. You could just as well use a large cardboard box for small numbers of chicks or a brooder in which the chicks are raised on wire. In any event, the basic requirements of any set-up are adequate space, ample feed and water, and strict sanitation.

Good broilers reach market weight (three pounds) in nine to 10 weeks. Layers can go to the laying house at this time, with at least four square feet of space per bird. This means that a flock of a dozen, ample for most homesteads, can be kept in a house six by eight feet.

The Henhouse

A concrete floor is nice to have in a henhouse, but it isn't likely to deter rats and mice unless the building also has a stone foundation, which gets a bit pretentious for a six-by-eight building. A wooden floor provides an ideal breeding place for rodents. Raising the building far enough off the

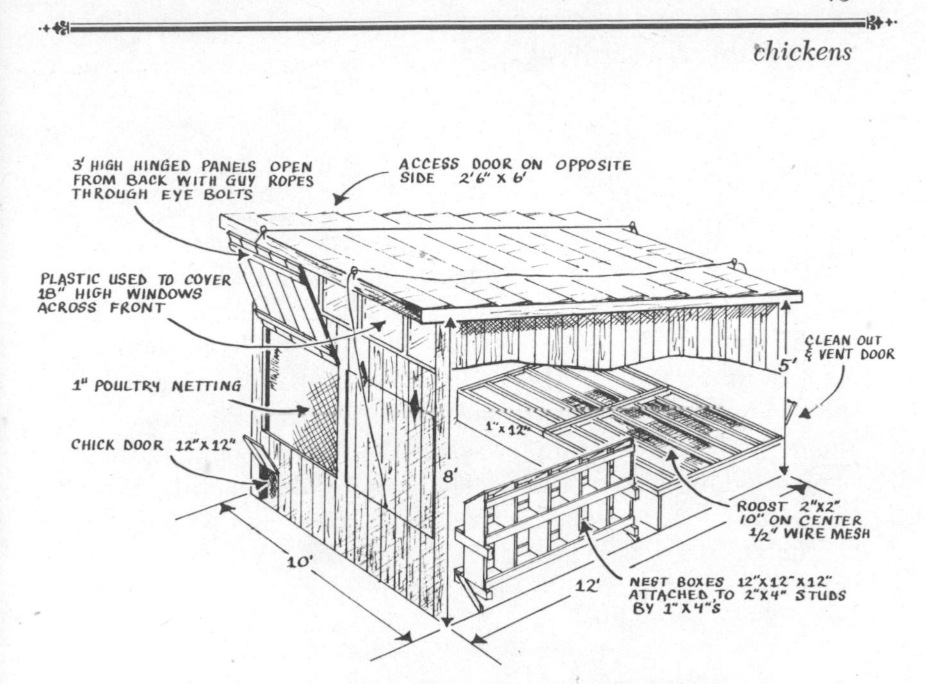

The chicken house pictured here is designed to provide plenty of ventilation. Notice that the windows open from the outside so they do not take up valuable space inside the building.

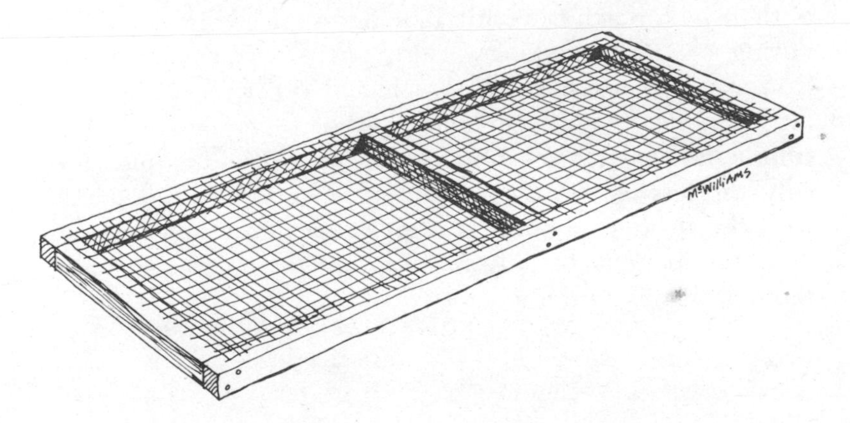

This is a detail of the false bottom that goes inside the nest box. The screening should be half-inch wire mesh.

ground so light and air can get under it will help considerably.

The biggest consideration in a poultry building is ventilation. A hen gives off a tremendous amount of moisture which must be removed from the building. This presents a special problem in cold weather. Ideally, windows should open from the top. The best window is probably one which slides down and can be opened or closed from the outside. I once built a much simpler system by merely hinging my scrap windows at the bottom. They could be opened by tilting them in. Although these windows were easy to install, ventilation wasn't as good as it is with windows that slide open from the top. What's more, when opened and tilted in, these windows took up valuable inside space.

WHAT CAN GO WRONG?

Chickens are heir to so many diseases it's a wonder any of them ever reach maturity. Reading most books on poultry raising is more like taking a short course in veterinary medicine than a course in chicken raising. Most homesteads have far fewer health problems with any of their livestock than commercial enterprises. One of the big reasons is undoubtedly the less crowded conditions on the homestead, along with the smaller numbers of a given species. Crowding and mass production not only invite problems, they also make those problems harder to control.

Cannibalism

In all the years I've raised chickens, I've only had one real, persistent problem: cannibalism. It usually starts about the time the pin feathers begin to appear on young birds. There can be many causes, but once started it becomes a bad habit and as a habit is difficult to control even if the cause is removed. Some of the more common causes of cannibalism are crowding, boredom, too much heat, too much light, and improper diet. Birds that are allowed to roam, which is the ideal of the organic chicken farmer, obviously are less affected by any of these factors.

The whole problem starts when one chicken starts picking at another, usually in the vent region. She draws blood, and pretty soon everybody's picking at the poor thing. It won't take long for the flock to kill her and still keep on picking until all that's left is a rather ghastly mess. And then they'll start on another one.

Many chicks are sold as "debeaked" which means a portion of the beak has been snipped off with a special hot blade which cuts and cauterizes at the same time. The debeaked chickens we've had have never turned cannibalistic.

We have found, however, that the best solution to cannibalism is pine tar. At the first sign of picking, we spread pine tar on the affected area. There is even a blood-colored pine tar made for the purpose, but the black works just as well. The birds keep on picking but find out the taste is awful, and they soon find some other way to amuse themselves. But you have to catch the picking early for this treatment to be effective.

Disease Control

I don't have any personal experience with any other major problems or diseases in chickens, and judging from the letters I get from hundreds of other homesteaders, neither do they. This I attribute to the conditions under which we raise the birds: adequate space, proper diet, and good sanitation. This is in sharp contrast to the commercial growers who medicate both feed and water as a matter of course because they know they'll have trouble if they don't.

One other factor that's important in disease control is an "all out—all in" policy. You don't put new layers in the same house with older birds. The used-up layers are butchered, and everything is carefully cleaned, disinfected, and aired out before the new batch comes in. However, I've violated even this basic rule with no problems. After all, Mother Nature doesn't sweep off the earth after each generation.

If disease should strike your home flock, the best thing to do is to immediately isolate any affected birds, then contact your county agent for advice. Health problems seem to be minimal in small flocks raised organically. You'll have mortalities, but that doesn't mean an epidemic. Dispose of any sick or dead birds immediately to avoid spreading any disease.

CHICKEN FEED

Feeding organically grown grains, once again, is one of the major hurdles for the organic homesteader when it comes to his poultry flock. While it is possible to get commercial feeds without medication or additives, some people want the added assurance that the feed their animals get has been grown on organically fertile soil, without chemical insecticides, herbicides, or fertilizers. In most areas today this is still a big order, unless you can grow your own. However, the homesteader who has a little room for planting will fare far better with chickens than with larger stock, if only because a six-pound chicken eats considerably less than a 500-pound sow!

Growing Your Own

Corn, for example, can be grown on a small plot. Small quantities of corn are easily harvested and husked by hand, and hand shellers are available at low cost. The amount of labor involved might not be tolerable for stock such as hogs or goats, but for some people, at least, it could be recreation in connection with a few chickens.

Some grains, such as millet, can be stored without threshing. During the winter, tie a small bunch of the stalks head down from the rafters in the henhouse, just high enough so the birds have to stretch to reach the grain. Exercise, and good food, too.

Sunflower seeds can be removed from the heads by rubbing them over a coarse screen. Fed whole to the chickens, they are an excellent source of protein.

Many other crops grown on most homesteads are acceptable chicken feed, along with much garden "waste" and clean kitchen refuse: vegetable trimmings, dry bread, and don't forget crushed egg shells as an important source of calcium.

We've heard of people who, during the Depression when they couldn't afford chicken feed, kept a close watch on highways for wildlife fatalities. The possums, squirrels, gophers, and such killed by cars were brought home for the chickens when meat scraps weren't available.

Free-range chickens make up a good portion of their animal protein needs from insects and worms, which is a graphic example of how organic farming works. The chickens not only save you money by finding their own protein (the most expensive ingredient of a feed ration), but they also help keep down the insect population at the same time.

Feed Rations

Here are two rations suggested by *Organic Gardening & Farming* which can be compounded on the homestead:

Starter and Growing Mash—about 20% protein

45 lbs. shelled corn
15 lbs. oats
10 lbs. soybeans
12 lbs. barley
2 lbs. fish meal
7 lbs. alfalfa meal
2 lbs. meat scraps
2 lbs. wheat germ
2 lbs. brewers yeast
1 lb. bonemeal
2 lbs. kelp

Laying Mash

40 lbs. shelled corn
15 lbs. oats
15 lbs. barley
10 lbs. soybeans
4 lbs. alfalfa meal or dry hay
2 lbs. wheat germ
1 lb. charcoal
2 lbs. meat scraps
2 lbs. bonemeal
1 lb. ground limestone
2 lbs. linseed meal
2 lbs. kelp
2 lbs. fish meal
2 lbs. brewers yeast

Mash is feed that is ground, and grinding it may present another possible problem for the small place. Very small amounts can be ground in small hand grinders, but it's a time-consuming job—one that will effectively increase the cost of your home-grown meat and eggs. But a farm-sized grinder—at least a new one—will cost thousands of dollars. A feed mill will do the grinding for you, if you have a reasonable amount ground at one time.

There are two main reasons for grinding feed. It increases feed efficiency, for one thing, but of more importance to the homesteader, it keeps an animal from picking out what it *wants* to eat at the expense of what it *should* eat, much as a child eats the pickles and neglects the carrots or spinach.

Scratch feed is not ground, but as you can see from the following suggested formula, it doesn't contain all the minerals a chicken needs.

Scratch Feed

35 lbs. cracked corn
25 lbs. oats
25 lbs. barley
15 lbs. sunflower seeds

As in all phases of feeding, suggested formulas are only that. Substitutions are not only permissible, they're also the rule rather than the exception, because feed formulas must be based on geographic area, price, and availability of various ingredients. The important thing is meeting the protein, vitamin, and mineral needs of animals of a given class and age.

Grit is an essential item for all birds of any age. It must be provided for confined birds, but ranging birds will seek their own. Grit is the small stones which end up in a toothless bird's gizzard and grind up its food. Be sure to use chick-size grit for chicks in brooders.

Another item of importance is oyster shell, which contains the calcium needed to manufacture egg shells. Grit and oyster shell are fed free-choice, that is, the birds have constant access to them from hoppers.

MANAGING YOUR LAYERS

Your chickens will lay their first eggs anywhere from four and one-half to six months after being hatched, de-

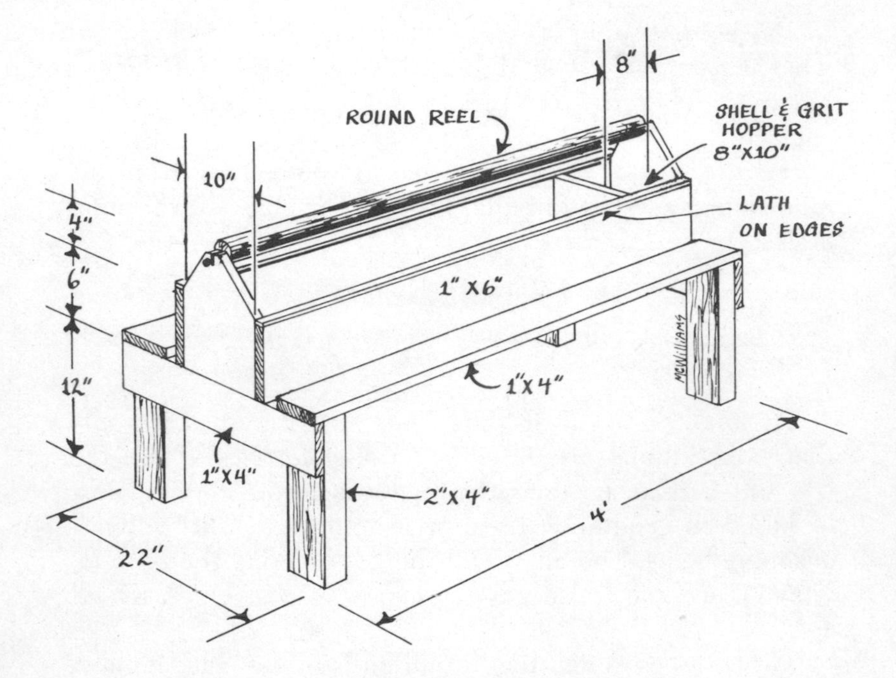

This simple feeder is ideal for layers, as it will keep grit and oyster shells separate from the feed so that hens can eat them free choice.

pending on breed and other factors. Our first pullet egg—no larger than a pigeon's—was the cause of much jubilation in our house, and I've since learned that this is the standard reaction. It's amazing how one tiny egg can make you feel so self-sufficient!

Egg Production

If you get chicks in the spring, they'll start laying in the fall and you'll soon be getting an egg from every bird almost every day. Visions of becoming an egg producer are likely to loom up at this time. But along about November or so, as the days get short, you'll most likely be in for a disappointment. Egg production will drop. The main reason is the decreasing length of day, which is more noticeable and therefore of more concern in the North. Spring is the natural time for birds to lay eggs, and that event is triggered by the increase in daylight hours. In the fall, the reverse is true. Science tells us that the pituitary gland, activated by light, controls egg laying.

Coupled with this (but of less importance), is the decrease in temperature. Cold weather doesn't bother chickens so long as they have warm or unfrozen water. However, they will have to eat more just to maintain themselves, so egg laying, which requires extra energy, is likely to suffer. When cold weather is accompanied by shorter days, with less time to eat and less stimulation of the pituitary, the effect of winter on egg production can be considerable.

The solution is to artificially lengthen the day with lights in the henhouse. You can turn them on before sunrise, or leave them on after sunset, or do a combination of both. The time isn't important, but the length of exposure to light is. Too much light, however, can cause problems such as cannibalism. A 14-hour day is optimum.

During cold weather, water should be checked more often than usual to be sure it hasn't frozen. Eggs, too, should be gathered several times a day to prevent them from freezing. (They should also be gathered frequently in very warm weather.)

As your flock nears its first birthday, you'll have to make a management decision. Is it time to get new chicks to replace those whose production will drop markedly when they reach the end of their year of laying five-to-six months from now, or will you carry over the old birds for an extra year?

How Long to Keep Your Layers?

Commercial egg producers invariably replace their layers annually because the rate of lay the second year is too low to be profitable. However, a commercial farmer's idea of profit might not be your idea of profit, and given the difference in management techniques, carrying over older birds might well be profitable for the homesteader.

Day-old chicks, for example, might cost 60¢ each in small lots (and much more for the more unusual breeds). Feed costs are highly variable, but you'll surely have several dollars per bird invested in feed before you get that first pullet egg.

Balancing those figures against the reduced rate of lay for second-year birds and the cost of feed during moulting when laying rate is zero, against the cost of raising replacements, you might decide to give the old gals another year. There is also the labor of raising chicks to consider, and the fact that, while the eggs of old birds are less numerous, they are larger. We even hear from folks who keep layers for a third year, and are satisfied.

It's possible to start raising chickens by buying birds that egg farmers consider all used up. Since these farmers seldom keep birds for the second year, they ship them to soup factories. The price a farmer gets is far below the cost of day-old chicks—to say nothing of feeding those chicks until they start laying. Chances are, if there's a commercial laying operation near you, you could get some of these birds at a very reasonable price.

Force Moulting

If these new birds aren't laying, or when your own birds stop laying, you might want to force moult them. Moulting is natural, and if you're out of eggs very long you like to sort of help Mother Nature along.

Here's how to force moult: Enclose the birds in the henhouse, and block out all light. Provide water, but no feed, for three days. After that, provide just a bulky feed, such as oats. The birds' feathers will fall out. After two weeks, resume normal feeding. When their new feathers have grown, the birds will resume laying.

Culling

An important part of an efficient operation, especially where older hens are involved, is culling. A four and one-half–pound bird eats about 60 pounds a year even if she doesn't lay an egg. (Feed consumption goes up with production. This means that a four and one-half–pound bird laying 200 eggs a year will eat about 89 pounds of feed.)

There are several important signs that show if a hen is a boarder. A young hen who hasn't laid an egg, for example, has a very yellow beak and feet. The yellow bleaches out as she lays, so if you have a hen who still has yellow feet when her sisters don't, she's a candidate for the soup pot. Laying hens have moist, round vents, while boarders' vents are dry and puckered. Layers have bright combs while culls' are dull and shriveled. A layer's pubic bones are farther apart than a nonlayer's.

Culls can be used in chicken soup, chicken and dumplings, and many other dishes where the meat is moist-cooked for any length of time.

POULTRY BUTCHERING

There are several different methods of slaughtering and dressing poultry . . . and as many variations of those methods as there are homesteaders. In general, any of the methods can be used for any type of bird.

The first method that comes to mind, and the one that is probably easiest for most people, is to hold the bird by the legs with its head stretched out over a chopping block, and to dispatch it with the hatchet. The headless bird is tossed aside to thrash around, or placed under a box or basket to somewhat restrain the flopping.

There are better ways. The more sophisticated homestead slaughtering operation will have a sticking knife, a narrow-bladed tool that can be used to advantage on any bird and is almost a necessity for small ones such as squabs.

The bird is hung by a cord attached to the legs. The knife is used to slash the jugular vein along the side of the head just at the top of the neck. The thin blade is then quickly inserted into the mouth and thrust through the roof of the mouth to pierce the brain which is located near the back of the head. This instantly loosens the feathers and makes plucking much easier. This is the favored method of killing.

Still another way to get the job done, much used in some European countries, is promoted by homestead women who have to do the butchering themselves but who don't feel comfortable with an ax. The bird is held by the legs with its head and neck on the ground. A stout pole (try a broom handle) is laid over the neck. Hold the pole down with one foot on each side of the bird's head and firmly pull up on the feet. The neck will still have to be cut to permit bleeding.

Some years ago a small guillotine affair was available. The bird's head was placed into it, the top was pressed down, the head fell into a waiting basket to be buried in the compost bin and the headless carcass could be placed under a box to bleed out. I've seen these at farm sales but haven't been able to find any new. An enterprising craftsman could probably make one.

Or maybe you're like the doughty old farmer's wife who simply went to the chicken yard, grabbed Sunday dinner by the neck, and spun it around as she walked to the kitchen to pluck it.

Decide which method you'll feel most comfortable with. Then decide whether you'll dry pluck, or scald. Dry plucking, as the term implies, is simply pulling out the feathers, dry. The sticking knife method of killing is almost a must for this, as piercing the brain loosens the feathers considerably. Dry plucking is said to produce better quality meat, at least in appearance, but scalding is much easier.

For scalding, you need a container of water large enough to dip a bird in. We use a 20-quart canning kettle. The water should be 126° F.: much cooler and it won't do the job, much hotter and it will damage the skin. Swish the bird in the water for about 30 seconds.

Then hang the bird by the feet or lay it on a clean table. Pull out the wing and tail feathers. Next work on the breast, the back of the wings, and the legs. The legs are left until last because the blood drains from the legs last and this will improve the appearance.

The faster you work the easier the feathers will come out. If necessary, give the bird another dousing in the hot water.

When the main feathers are removed, get out the pin feathers. If you're doing several birds at one time the pin feathers can be left until last as they'll come out somewhat easier if the bird is cooled and dry. Grasp them between the edge of a blunt knife and your thumb, or use any of the various kinds of tweezers made for the purpose.

Poultry can be cooled and then drawn, or it can be eviscerated immediately. Normally, chickens are gutted at once while geese and turkeys are allowed to ripen for a day. Ideal temperature for this aging is 40 degrees. They can be cooled in ice water, although they should not be allowed to soak for more than 20 minutes, again mainly for appearance.

The easiest way to clean a chicken or other fowl is to lay it on its back, hold it firmly with your left hand (if you're right-handed) on the breastbone, and using a butcher knife, make a horizontal cut just under the keel. All the insides can be removed through this opening by reaching in and pulling. You'll have to cut around the vent to remove it, and the crop and windpipe require quite a bit of tugging.

A more refined method is to cut the skin along the back of the neck and to remove the crop and windpipe through this incision. The neck can then be cut near the shoulder inside the skin, leaving the neck skin to be folded over the back during roasting to prevent the meat from drying out.

Next make a cut below the back of the breastbone. Then make a cut completely around the vent, taking care not to cut the intestine by inserting your finger into the first cut to maneuver the intestine out of the way. Push the vent and intestine into the body and pull them out through the first cut. All the intestines are then removed through the first opening.

Take care in cutting the gall bladder from the liver; cut off a piece of the liver with the sac if necessary. Slice the gizzard lengthwise, remove the contents, and peel off the lining.

All waste goes into the compost bin, well buried so dogs and predators can't get at it. And your bird is ready for the pot or freezer.

TURKEYS

A DICKENS TO RAISE

Of all homestead poultry projects, turkey-raising seems to appeal to the least number of people. Turkeys are among the most difficult domestic fowl to raise. They are amazingly stupid—from the newly hatched poults who can starve to death while trampling in their feed because they haven't learned where to find it, to the hens who lay their eggs standing up. (Some breeders use special rubber mats in the nests to help cushion that drop.) They are easily frightened. An acquaintance of mine who raised turkeys commercially went wild every Fourth of July because the fireworks in a nearby village invariably sent thousands of birds piling up in corners where they'd suffocate unless he waded in and unpiled them. Airplanes going overhead had the same effect, and they didn't care much for thunder, either. And turkeys are much more susceptible to disease than other fowl are, especially if raised around chickens.

But if a home-raised, golden-brown, juicy turkey with rich dressing and thick gravy appeals to you, go ahead and raise your own.

BREEDS

The turkey is a true American bird, although the breeds available today bear little resemblance to the native specimens hunted by the Indians and Pilgrims. As is the case with all other domestic animals, selective breeding has produced "new" stock designed to meet specific needs. Much of the early selective breeding of turkeys was done in Europe, strangely enough, to produce a bird with shorter legs and plumper breasts, resulting in more meat per bird. Later the white breeds became popular (white poultry of any kind is easier to dress) and still later smaller breeds were developed which helped promote this as an "everyday" meat.

The Bronze, which school children still color at Thanksgiving, has largely been replaced by the less spectacular White Holland and the smaller Beltsville White. There are a number of other breeds, but since these three are of some commercial importance they're probably the ones you'll be able to find.

Half a dozen to a dozen birds should be sufficient for most families. You'll start with poults (the turkey equivalent of a chick), probably ordered from ads in the farm magazines.

THE BROODING PERIOD

Brooding equipment is much the same as that used for chickens. However, if you use *any* chicken equipment for your turkeys, be sure to disinfect it. Thoroughly clean it with hot, soapy water and a stiff brush, then disinfect it with one ounce of lye to one gallon of water or with any good commercial disinfectant.

Most homestead poults are started in early summer when warm weather has pretty well settled in. In such cases, brooding facilities in a battery should be provided for about 10 days. If no battery is available, a box about 20 by 24 by 15 inches high with a 100-watt light bulb inside will work.

One of the first chores you'll have is teaching the new poults to eat. One way to get them to eat is to sprinkle chick scratch on top of the ground turkey starter mash. The coarser scratch—usually a combination of cracked corn, wheat, oats or other grains depending on local availability—seems to attract the birds' attention more readily than just the mash, and they're more inclined to peck at it. As they learn to eat, the scratch is eliminated.

THE SUNPORCH

After the brooding period the young turkeys go to their sunporch. In spite of the common belief that turkeys cannot be raised on the same place with chickens, it *is* possible. The secret is to keep the turkeys in cages raised off the ground, on sunporches.

One of our neighbors customarily raises six to 12 turkeys a year right next to his hen house, in a pen about five feet wide, 12 feet long and about two feet high. The entire affair is raised about three feet off the ground. About half of the pen is roofed over to protect the birds from rain and direct sunlight, and roosts are provided. Each bird requires about five square feet of space.

Floors can be made of 1½-inch mesh made of heavy gauge wire. Supports made of wire attached to turnbuckles can be kept taut and will prevent the floor from sagging.

Another type of floor, perhaps more to the liking of

Keeping turkeys housed in structures several feet off the ground will prevent them from picking up many diseases to which they are susceptible. Sunporches need not be as elaborate as the one pictured here.

homesteaders, can be made of 1½-inch square strips of lumber spaced 1½ inches apart. In fact, if you have more old lumber than wire or money laying around, as most of us homesteaders do, the sides as well as the floor can be constructed of wood. Make these of vertical lath spaced one inch apart.

WATERING AND FEEDING

For water, you can use regular poultry fountains. (Again, don't forget the thorough cleaning and disinfecting if the fountain has previously been used for chickens.) The fountain will have to be placed inside the pen and removed for filling and cleaning.

A simpler method for providing water for a few birds is to cut a hole in the side of the pen large enough for any pan you might have, and to fence it in with heavy gauge wire spaced three inches apart at the bottom. The wires are brought together at the top and fastened to the side of the pen so the arrangement looks like half a bird cage. This way, the pan can be filled and cleaned from outside.

The feeders, similarly, can be regular chicken feeders that will fit inside the pen, or a simply constructed wooden trough that can be filled from outside. Obviously, the feed should be protected from rain. Two inches of feeding space per bird is recommended.

It takes about four pounds of feed to grow a pound of turkey. For the home flock, so little feed will be used that it will hardly pay to mix the meat scraps, minerals, and other ingredients needed for a balanced ration. It will be more economical to buy prepared feed. Pellets for feeding turkeys are available from several companies, but read the label carefully: many of these feeds are medicated.

Corn tops the list of grains fed to turkeys for fattening. Oats can also be fed, especially if cannibalism or feather picking is a problem, since the high fiber content of this grain is generally recognized as one means of reducing feather picking (in chickens as well as in turkeys). Other grains, notably sunflower seeds, are also good for turkeys.

In addition, the homesteader will want to use liberal amounts of green feed. In fact, if possible, turkeys can be raised on range with a great saving in feed. If you don't have ground that chickens have never been on, or ranging chickens (and what homestead doesn't have at least a few Bantams running around?) it's best to leave the turkeys on the sunporch and to bring the greens to them. Among the best greens that can be grown on the small place for turkeys, or chickens for that matter, is Swiss Chard. As the outer leaves are cut, new ones grow. And they continue to grow until hard frost.

Rape and alfalfa, as well as lettuce, cabbage, and most any other garden greens, all provide good food for turkeys. As much as 25 percent of the ration can be greens . . . which can enable the homesteader to compete price-wise with the commercial grower.

The turkey pen is another place to make good use of excess milk from your goat herd. The milk, or skim milk or whey, should be used to moisten mash. Be careful not to provide too much so that it isn't cleaned up promptly as it will ferment in the feeders, attracting flies and becoming generally unsanitary.

Turkeys grow most rapidly during about the first 24 weeks. If feed prices are high, it becomes less profitable to hold them much beyond this age. Turkeys require "finishing" before slaughtering, especially if they have had a lot of greens in their ration. Corn is the most common finishing grain, but turkeys won't eat much corn before cool weather sets in in the fall, so finishing before then can be difficult.

DISEASES

Turkeys are notoriously disease-prone, particularly to Blackhead. This is an organism hosted by the small roundworm of the chicken. Keeping the two birds separate, even to the point of never walking from the henhouse to the turkey yard, will go a long ways toward controlling this disease. Leave a pair of overshoes at the turkey yard to wear when working with them, and only when working with them. The sunporch will eliminate this nuisance.

Birds affected with Blackhead will be droopy and the droppings will be yellow. If the birds die and are autopsied, the liver will frequently have yellowish or whiteish areas. One of the remedies used by commercial growers is phenothiazine. However, taking steps to prevent the disease, such as having a raised sunporch, is a more acceptable control measure for organic homesteaders.

Coccidiosis, while not as prevalent among turkeys as it is among chickens, is another problem you should be on the alert for. The usual symptom is blood in the droppings, as well as a general unthrifty appearance. Since wet litter is one of the predisposing factors, keeping the litter of young poults dry by frequent cleaning and by using heat (the light bulb) in damp weather, and by using sunporches off the ground for older birds, will help control this problem.

Pullorum is no longer the problem it used to be in both chickens and turkeys due to the rigid inspection programs carried out now at most hatcheries. It's good insurance to buy from a reputable hatchery where birds are U.S. Pullorum Clean.

Paratyphoid is less easily controlled, since carriers cannot be removed from the breeding flock as with Pullorum. Birds infected with this disease usually develop a greenish diarrhea. Losses of 50 percent and more can occur. There is no effective control.

Crop bound is another turkey problem, usually brought on by eating litter or green feed that is too coarse, such as cabbage. A heavy, pendulous crop results. The bird is still edible and should be slaughtered even if not fully mature.

For control of these and other disease problems that might strike your flock, check with your county agent. As with any other bird or animal, the best insurance is to start with good stock, provide ample room and proper nutrition, plenty of clean water, and maintain strict sanitation practices.

BIRDS OF DIFFERENT FEATHERS

In addition to chickens, there is a great deal of interest among homesteaders in geese, guinea fowl, ducks, and to a lesser degree in pigeons, game birds, and other fowl. While none of these are of much commercial importance, any or all of them have a place on the homestead. The only deciding factor, actually, is whether you, personally, are interested in raising them.

GEESE

Geese enjoy wide popularity because of their size, which makes them something of a festive treat for the homestead table, and because the flavor of the meat is different from the chicken and rabbit which are likely to be homestead staples.

In addition, these majestic birds are just plain fun to have around, and of great importance to the serious homesteader, they are very easy and cheap to raise.

The three main breeds of geese in this country are the white Embden, the grey Toulouse, and the white Chinese. Of the minor breeds, the Pilgrim seems to be mentioned

These beautiful birds are cheap and easy to raise on the homestead because they will forage for themselves after they are a few weeks old.

quite often among homesteaders. Embdens usually weigh about 12-14 pounds, the Toulouse is similar in size, and the Chinese and Pilgrims are smaller.

Day-old goslings are rather expensive. Expect to pay $1.50 or more. They can be started out like chickens so far as the brooding is concerned. Commercial duck or goose starter feed is available, but never use commercial chick starter for goslings or ducklings, as they'll be killed by many of the formulations. For the organic homesteader, a good starter ration is dry bread crumbs moistened with goat milk.

Geese don't need water for swimming, but they drink great quantities of it and they drink by immersing their entire heads, so the water holder should be fairly deep. An ordinary water bucket works fine. It gets quite dirty and must be changed often. Very young goslings shouldn't even get wet from dew-laden grass before they've feathered out, much less attempt to swim.

If greens are available, provide them as soon as the young geese will eat them. Within a few weeks grass will be their entire diet, which means as long as you have a place to let them run, your feed bill stops. There's no cheaper way to put meat on your table!

Many people are interested in using geese as weeders. My own experience with this has not been happy. A goose in the garden, for me, has been like a fox in the henhouse. Geese are widely used to weed crops such as cotton. They'll weed strawberries (before the berries are ripe, after which they'll eat the berries), but in the average garden they'll clean up the corn, beans, lettuce, and everything else along with (or instead of) the weeds. Better plan on fencing off the garden, or the geese.

In the fall, as the grass starts to dry up, geese are fattened with grain, especially corn. Geese hatched in the spring are butchered that fall. Older birds aren't exactly the delicacy you dream of serving for Christmas dinner.

Goose butchering is similar to chicken butchering, with one major exception: they are a dickens to pluck. Plucking a large goose is just about an all-day job. People who regularly butcher geese have developed a few techniques to ease the job. For one thing, the hot water used to loosen feathers doesn't penetrate the waterproof oil which coats duck and goose feathers. Some people add detergent to the water to help overcome this. Others say shampoo rinses off more easily and works better.

After the main feathers are removed, coat the entire goose (or duck) with wax. It takes an awesome quantity of wax and a witch's cauldron to dunk a goose, but I've found that simply pouring melted wax on the feathers works quite well. The wax is peeled off with the fingers, and the down comes off with it.

Down, of course, is a valuable commodity on many homesteads, and you may not want to ruin it with wax. Hang the down and small feathers in bags to dry. The down is simply stuffed into pillows or quilts. (We haven't yet heard of a way to keep the tiny feathers from floating all over: don't try this in the living room, or certainly not where there's a breeze!)

Some old-timers swear that duck and goose must be aged, much like beef, to bring out the flavor. It's hung in a place safe from rodents and insects, for a week or more depending on temperature and other factors. An old refrigerator with screened ventilation holes, situated in a cool place, is ideal for aging birds.

No doubt you'll want to save a few geese as breeders, so make sure you don't butcher all the ganders (or all the geese). This is easier said than done. The way to tell the difference between the sexes is said to be by placing the goose on its back and pressing gently, but firmly, with the genital region between your index and middle fingers. The goose will show a round opening, while the gander's will be less round. I must confess, though, that this hasn't worked for me. Since both geese and ganders hiss, and both are very similar in appearance and other characteristics, sexing geese is difficult for the inexperienced. You'll know if you guessed right along about February, when they start laying.

Goose eggs take 28 days to hatch, and one goose can cover about 12 eggs. Some breeders get more eggs by removing the first ones as they're laid, and by feeding the geese the high-energy laying mash described in the chicken feeding chapter. If you do this you'll need an incubator to help the geese hatch all their eggs, although a broody hen can hatch goose eggs too. When the goslings hatch, you're off on the cycle again.

If you discover you don't have a gander, or want to replace any bird with another mature one, expect trouble. Geese mate for life and are very clannish, and in most cases, introducing a new member to the flock takes some time and patience.

DUCKS

In the duck family, White Pekins are by far the most popular meat breed. The famous Long Island ducklings are Pekins. But homesteaders are a diverse lot, and almost any breed can be found on small farms.

Some people prefer duck eggs to chicken eggs. Most ducks are seasonal layers, so if duck eggs interest you consider a breed such as the Kahki Campbell or the India Runner. These lay as well as some chickens. As with hens that lay at a high rate, these breeds are not valued for meat because of their small size and slow growth. Kahki Campbells, with proper housing and feeding, have laid as many as 335 eggs per year. Like the Leghorn hen, they don't make good setters because they're rather nervous.

Since most ducks lay in the early morning, they can be

penned up at night and released in mid-morning to forage. They will eat a great deal of grass if allowed free range. Nests should be 12 by 18 by 12 inches deep.

For meat, whitefeathered fowl of any variety are usually preferred because they dress out more attractively. In ducks, this means the White Pekin and the White Muscovy.

Other breeds are used by homesteaders, of course. These include the Muscovy, Mallard, Rouen, and others that are less common.

Pekins, which lay about 160 eggs a year under proper management, weigh about eight to nine pounds at maturity. Hens are somewhat smaller. They are usually butchered at about six pounds (much more tender then, and more efficient feed conversion) which should be at about nine weeks of age. Some breeders claim their strains reach seven pounds in seven weeks on 22 pounds of feed. . .showing once again the importance of genetics, even to homesteaders.

Muscovies are somewhat larger. They lay about 40 to 50 eggs a year.

GUINEA FOWL

Guinea fowl are very popular among homesteaders. Like geese, guineas are good watchdogs (although unlike geese their bark is worse than their bite). The meat of young birds is tender and reminiscent of wild game such as grouse, partridge, or quail. . .pretty elegant stuff in these days of high prices for even hot dogs! But the homesteader can manage it quite easily.

There are Pearl, White, and Lavender guinea fowl in the United States, although there are many other species in their native Africa.

Guineas will cross with chickens, but the offspring will be "mules," or sterile. Guineas eat bugs like chickens do, but they don't scratch, so they are much less destructive in the garden.

Guinea chicks (which are called keets) may be purchased by mail just as chickens are. The brooding is similar to that for chickens, which has already been described.

Guinea fowl are almost like wild birds: they require little care or purchased feed, and make elegant dining.

Keets are more delicate, or maybe just dumber, than chicks. They often drown in even the small chick-size waterers, so you'll have to make the water more shallow by putting pebbles in it.

If you raise keets from your own guinea eggs, it's even worse because guineas are lousy mothers. They don't have enough sense to come in out of the rain or at night, which is tough on the little ones. A chicken can be used to hatch guinea eggs, and makes a much better mother.

One of the biggest problems guinea raisers seem to face is sexing their breeding stock. According to the books, the hens say "buckwheat" while the cocks have only a one-syllable shriek. The problem is, the hens also use that one-syllable shriek. The male and female look almost exactly alike except that the male has a somewhat larger helmet and a coarser head and has wattles.

The best way to sex the birds is to close them all in the coop. Then go in with a sack, and grab everything that says "buckwheat." When the sack is full and there are so many "buckwheats" you can't tell who's saying what, take it out and start over with a fresh sack. When you've taken out all the birds that say buckwheat, you should have only males left in the house.

A good guinea hen will lay about 100 eggs a year, but on most farms only a fraction of these are hatched—and most of the keets are lost. One reason is that these birds are almost wild: they much prefer to live in trees than in henhouses, and eat bugs and weed seeds rather than the nice grain you put out for them.

Their preferences make them fine for the organic homestead, but make it difficult to find their nests and follow sound management practices. If the nest is left undisturbed the hen will lay about a dozen eggs and then go broody. But if you find the nest and remove some of the eggs to place under a hen or in an incubator, she'll continue to lay for a much longer period. Don't take all the eggs: mark two of them so you don't steal them on subsequent raids, and leave them. If you take all of them, the hen will find a new nest, and the hunt will begin all over again.

The eggs can be eaten, of course, as can all eggs. They are smaller than chicken eggs, with extremely hard shells.

PIGEONS

Pigeons are almost as good as guineas when it comes to providing free food for the homestead with little labor. Pigeons are as common as sparrows around most barns, and all you have to do is harvest the squabs just before they leave the nest. Crawling around the rafters of a barn in the dark with a bag and a flashlight might not be your idea of a good time, though.

So fix up a loft. Confine the birds for a few months and provide water and grains such as cracked corn, milo, Canada peas, and wheat. In a few months, you can let the birds roam again—but they'll continue coming back to the loft. No feeding, no watering, but you'll get the guano for the garden and the squabs for the table.

Pigeons can be raised in confinement, of course, in which case such breeds as the White King, Giant Homer, or Giant Runt will be a wiser choice than common barn pigeons. Since pigeons are monagamous, and lay only two eggs in a clutch, confined birds and purchased feed will make this one of the more expensive homestead meats.

Pigeons mate for life at about six months of age. The eggs take about 18 days to hatch and the squabs are ready in about a month—whenever they're feathered out under the wings, and before they leave the nest.

Most pigeon raisers are fanciers, which means that there's a lot more information on pigeons available than there is on such fowl as guineas. As a matter of fact, my own livestock raising experiences started with pigeons, when I was about 10 years old. We still have several dozen rollers—birds that turn somersaults as they fly—and I must confess that I enjoy them so much I have never eaten a pigeon. But there's more to homesteading than just putting food on the table, isn't there?

Most of the other poultry raised on homesteads falls into this classification, although, with experience, proper management and the right situation, it may be possible to make a little cash income from certain species. Everything from partridge to pea fowl can be raised for fun, but they can hardly be considered part of the self-sustenance aspect of homesteading.

GOATS

THE BETTER DAIRY ANIMAL

Just as dairy products are one of the groups of basic foods necessary for good health, dairy goats are important for the "health" of the self-sufficient homestead.

From the technical standpoint, it's important to mention that goats give more "useable" milk than cows; that goats are more efficient milk producers; that they are easier than cows to handle; that goats do not get tuberculosis so drinking raw milk is less dangerous if you have goats; and that goats are less expensive than cows and therefore easier to get started with. They obviously require less work and housing space, you can take a doe to the buck for breeding in the trunk of the car, and the milk they produce is more easily digested than cow milk.

Those are the rational reasons for raising goats. But once you get to know a goat personally, those rational reasons fade into insignificance! Chances are, you'd want to keep a few goats even if they weren't the most utilitarian, efficient, and economic animals on the place. They're friendly, intelligent, and full of personality and fun.

People either hate goats, or they love them. The hate is hard to understand, unless it stems from the mass of myths that abound concerning goats. Most people have never even seen a goat, yet they "know all about them," from comic strips.

Myth number one is probably that goats smell. While this may have had its origin in the aroma of the buck during the mating season (roughly from September through January), this is of small concern to the homesteader. In the first place, most people do not keep their own buck, for reasons we'll look at when we talk in more detail about the herd sire. The does do not smell. In fact, a goat is one of the most fastidious of creatures, and if given an opportunity to keep herself clean, she will be less objectionable than most dogs.

Everybody also knows that "goats will eat anything, including tin cans." Ridiculous. Maybe one of the problems here is that goats are quite unlike most other animals we're familiar with. Not only are they not carnivorous like cats and dogs, they aren't really grazers like sheep and cattle, either. They're more closely related to deer and other browsers.

What this means is not only that goats prefer trees and bushes to grass, but that they also take a bite here, a nibble there. And that includes your shirttail or clothes drying on the line, or anything else that looks interesting. As for tin cans, they will eat the paper off of them, but after all, paper is made from trees, and goats like trees.

Still another widely held false impression of special importance to the serious homesteader is that goat milk tastes "strong" or "goaty." There are several possible explanations for the rise of this myth. In the first place, it *is* true that *some* goats have off-flavor milk, but what most people don't realize is that some cows do too! Yet you never hear of "cowey" milk. We'll look at milk more closely later, but for now the point is that your homeproduced goat milk can be as delicious as it is nutritious.

Which brings us to another myth that, surprisingly enough, is in the goats' favor: the high butterfat content and nutritive value of goat milk. The composition of goat milk is very similar to that of cow milk. From the standpoint of nutrition, there isn't enough difference to shake a stick at. And most people can't tell the difference in taste, either.

Goat milk is somewhat more easily digested, and is very important in cases where infants, especially, are allergic to cow milk. But for all practical purposes the two milks are the same. Goat milk is not a medicine, but a food.

Likewise, goat milk is no richer than cow milk. There is a wide difference in butterfat content among individual cows, between breeds, and at different seasons of the year. Ditto for goats. The range is the same for both species.

Of course, cow milk bought in stores is "standardized," that is, some of the butterfat is removed and used in other dairy products. The usual level, set by law, is around 3.25 percent, and a dairy isn't going to make its milk richer than it has to because it would lose money. A homesteader, on the other hand, drinks the whole milk, whether it's from a cow or goat, and therefore gets a richer product.

It's a fact—not a myth—that more people in the world drink goat milk regularly than cow milk. In the United States, however, mass production and mechanized labor favor the less efficient cow. Goat dairying as a commercial enterprise has never flourished for many reasons, most of which center around the marketing of a product for which no great demand exists.

Goats were included in the 1970 agricultural census, but the figures are so inaccurate that they aren't worth reporting. The census takers counted 1,000 goats in Wisconsin, for example, and there are two herds that account for that many alone. There are a dozen people with goats that weren't counted right here in Marshall (population 1,200). Most goats, of course, are on homesteads rather than large farms, and homesteaders aren't invited to participate in the agricultural census.

In other words, no one really knows how many goats there are in this country. However, we do know that interest in goats is flourishing. *Countryside & Small Stock Journal* surveyed the extension service leaders in each state in 1972, and one of the questions asked was how many 4-H dairy goat projects there were today compared with five years earlier. Iowa reported "10 or 15 five years ago, but 150 today." California went from 600 to almost 1,600. And in the states that reported none, the magazine soon had letters from indignant local leaders to the contrary. All of this indicates that, at least for people intent on doing more with less, the dairy goat is coming into the respect and appreciation it so richly deserves.

GETTING YOUR GOAT

While goats are rapidly increasing in popularity, the goat population is spotty, and finding good animals is not always easy. It will pay to be armed with as much knowledge as possible before setting out in search of your back-yard dairy herd.

As you may have already noticed, we never speak of billy goats or nanny goats. I have no idea where these terms originated or if they were intended to be derisive when they were coined, but they're considered very bad taste today. Think of the most profane term anybody could use about your mother, and the indignation you feel will approximate what most goat breeders experience when somebody calls their herd sire a billy goat. Male and female goats are called bucks and does, respectively.

Selecting Stock

Stock less than a year old are referred to as yearlings, or sometimes doelings. Young goats, of course, are kids. Here are some other terms you're likely to run into:

Purebred stock is just what it sounds like—stock descended from animals that have been bred pure, without outside blood from other breeds or animals of unknown ancestry. Purebreds aren't necessarily registered (nor are they necessarily better than animals that aren't purebred).

Registered goats have papers. There are two registry associations in the United States. The American Goat Society (AGS) registers only purebreds, but the much larger and more popular American Dairy Goat Association (ADGA) also maintains herd books on grades, Americans, and experimentals.

A grade is a goat with one parent purebred and the other of unknown ancestry. An American is at least seven-eighths pure. To illustrate, if you bred a scrub goat (one of unknown ancestry) to a purebred buck, the offspring would be a grade with one-half purebred blood. If that offspring is bred to another purebred buck, the result would be a goat that was three-fourths pure. After one more go-round with a purebred buck, the result will be a goat that is seven-eighths pure. Obviously, it can never be 100 percent, so this is taken as an arbitrary cutoff point.

Experimentals most frequently result when a purebred buck jumps the fence and gets to a purebred doe of another breed. "Experiments" in the usual sense of the word don't work out and are frowned upon. A Saanen-Nubian cross, for example, doesn't produce a goat that gives lots of milk that scores high on a butterfat test.

For the homesteader, the fact that a goat is purebred or registered doesn't mean very much. Many animals are kept just because they're purebred, even though they have neither conformation nor production to recommend them. The only requirement for purebred registry is having purebred registered parents, even though they might have been worthless.

However, don't disregard these animals out of hand. The very finest animals are purebred. Registration papers are a valuable tool for the breeder who knows how to use them to improve his stock. Moreover, it costs the same to keep a purebred or a scrub. The initial cost will quickly be overshadowed by upkeep. Since the offspring of purebreds are obviously worth more, this is an important consideration where a market exists for breeding stock. You will almost certainly have goats to sell in a couple of years, and the extra income from purebred stock can go a long way toward paying the feed bill. Most small-scale breeders figure their stock sales are more important economically than their milk production.

A good grade may be better than a poor purebred, but the risk is greater. Since it takes experience to know the difference between a good goat and a poor goat, most people prefer to start out with cheaper stock, and work their way up. Price and availability are usually the deciding factors.

Breeds

Likewise, the choice of a breed is often a matter of personal preference and availability. There is no "best" breed, and a good doe of any breed is to be preferred to a poor doe of any breed.

There are five breeds of dairy goats commonly seen in the United States. Some are more popular in certain areas than others. Take this into consideration when making your choice, for not only will you want stud service from a purebred buck within easy driving distance, but it will be easier to dispose of extra stock if you have the breed that seems to be favored in your locality.

The Saanen is the Holstein of the goat world: she's noted for her ability to produce great quantities of milk, but with lower butterfat average. A new record was set in 1972 when a Saanen gave over 5,300 pounds of milk, or 2,650 quarts. Saanens are always all white, with erect ears.

Nubians are attractive to many people because of their long, drooping ears and Roman noses. While some Nubians are very good milkers, they are generally classed with Jersey cows, which are famous for high butterfat, but low production. Since butterfat is very important on the homestead, not only for giving milk its richness, but because butterfat is what makes butter and rich-tasting cheese, Nubians are popular among homesteaders. Nubians can be any color, or even spotted.

Alpines are divided into Swiss and French varieties, although the French are more common. These have erect ears

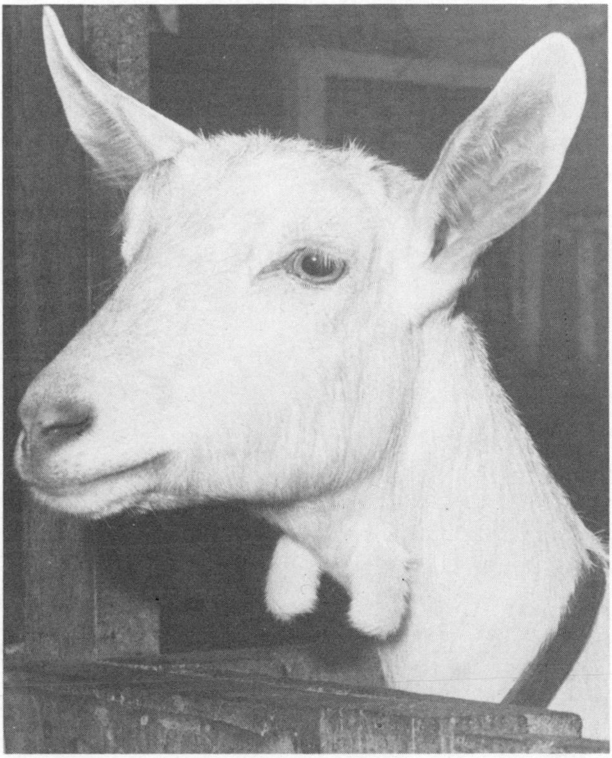

If you are interested in top-notch production, look for the Saanen which gives more milk, on the average, than any other breed of goats.

and may be any color, including several rather distinctive color patterns.

Toggenburgs, on the other hand, have only one color pattern. They are always brown, with white stripes on the face and white markings on the rump. They have erect ears.

The newest breed, and one that's growing fast in popularity, is the LaMancha. They are not purebred in the same sense as the others. Toggenburgs, for example, have been

Nubians are identified by their pendulous ears and Roman noses.

registered since the 17th century, which makes them the first registered animals in the world. But LaManchas can be registered with ADGA as Americans, which means they are seven-eighths "pure."

Their main distinction is their ears, or lack of them. Many people with a preconceived notion of what a goat

The more common variety of the Alpines is the French. They may be white, black, brown, gray, or combinations of these colors.

should look like are repulsed by LaManchas the first time they see them. But you don't milk the ears, and as a group, LaManchas are fine milkers. They produce a lot of milk which is high in butterfat. And not only do you become accustomed to the ears after awhile, you'll also find that LaManchas have completely different personalities. They are docile and sweet.

Any goat of any breed can have horns, be it a buck or a

Toggenburgs, originally from Switzerland, have very distinct white markings.

doe. Most breeders remove the horns within a few days after birth, before they start to grow, with either a disbudding iron or a caustic. There are also polled, or naturally hornless, goats in all breeds. Beards, likewise, are neither a sex nor a breed characteristic.

In addition to horns and beards, another decoration you're liable to wonder about are wattles. "What are they!" most people exclaim when they see them. Wattles are simply small decorative appendages of skin, most commonly found hanging on the neck. They don't do anything; they're just there and are completely natural. Toggenburgs are more likely to have wattles than the other breeds.

Shopping Around

Thus armed with basic terminology, you're ready to begin your search for a goat. Where do you look? While goat numbers are increasing, the population is still spotty, and if you're in an area where the animals are scarce you may have quite a chore ahead of you. The West Coast, the Midwest, and the Mideast all have sizeable goat populations and fairly active goat clubs.

If you have nowhere else to start, join one of the registry associations and get their membership lists. Chances are there will be someone listed who will at least be within driving distance.

Also check to see if there are any goat clubs in your area. Some clubs are so-so, while others do a great deal to help both beginners and more experienced breeders get more enjoyment and profit from their animals. Many have newsletters which include ads, and some even print "goats wanted" ads free to help newcomers get started.

Spring is generally the best time to buy a goat, or at least the easiest time. Milk is plentiful, and most goat barns are bursting at the seams with animals they weren't designed to house. Later in the summer and on into fall, many of the animals with short lactations will be dry and the demand for animals is higher. During the winter, a milking doe is worth her weight in imported cheese.

If you can visit several goat farms, you're fortunate. Look around, see how to do things (and in some cases, how *not* to do things). Ask questions.

Goat pricing is largely hit-or-miss and often highly irrational. Most people who raise goats simply don't know what their costs are. Others don't care, often because they consider goats a hobby. Still others simply get overstocked, and in an area where demand is light they may price the animals just to move them out.

Records show it costs about $70 to keep a goat her first year, before you get a drop of milk. Therefore, if she's good enough to keep at all, we could consider $70 a base price.

However, in a survey of experienced goat raisers, it was reported that young kids were sold for from $25 to $200. And more than once we've heard of goats being given away simply because somebody got tired of them. At the other extreme, two animals were sold for $1,300 each at the annual ADGA Spotlight Sale in 1972. Not enough goats are bought and sold to create a visible market value. It will probably pay to shop around.

There are various rules of thumb to determine how much you can afford to pay for a goat, but all of them are based on the amount of milk you'll get, and this is largely an unknown factor until you actually start milking the animal. One system that makes sense for homesteaders is to figure that a cow gives about six to eight times as much milk as a goat, so a goat should be worth about one-sixth to one-eighth as much as a cow, which has a more definite market value.

Should you get a kid, a yearling or a mature doe? There are advantages to each.

With the kid, you'll get to know each other before she freshens, and the initial cost will be lower. That doesn't mean this is the cheapest way to buy a goat, because you might go a year before getting any milk, and the cost of the goat by that time will probably be more than the purchase price of an older proven animal in the first place.

A doe, bred or open, purchased in the fall, will make a good adjustment before freshening. A breeder isn't likely to want to part with a doe as kidding time approaches, and if you have no experience with goats you'll feel more at ease if you get some before kidding time. A milking doe, of course, will provide milk immediately. She's likely to be more expensive than a dry doe.

Moving any animal is a stress factor. Even if you use the same feed as the previous owner (which is always a good idea for a few days—if you intend to feed something else, make the change gradually), the difference in environment is upsetting. Different housing, different fencing, different voices and faces, and such disturbing factors as other animals, all take their toll. The young or dry animal can get used to this environment gradually, but you can expect the stress of the milking animal to show up at the milk pail.

FURNISHINGS

The major requirement of goat housing is that it be dry and draft free. Goats are kept in everything from oversize dog houses to home basements. Minimum space requirements are about 20 square feet, assuming there is access to an exercise yard or pasture.

The roof, obviously, should be sound and watertight. The ceiling should be high enough so you don't clunk your head on it in late winter—the floor will be about two feet higher then, if you let the bedding accumulate. Insulation in the ceiling is a good idea in northern areas, but be sure it's high enough or protected enough so the goats can't get at it.

This suggestion applies to walls as well. Insulation is a

good idea, but whether the walls are insulated or not, they should be impervious to chewing, butting, rubbing and standing, all of which are favorite goat pastimes. Goats will put holes in half-inch plywood in a short time. Stout planks make good walls.

Floors are controversial. Most people keep goats in buildings that were designed for something else and simply accept the floor that came with them. Concrete is very nice. It's easy to keep clean and easy to get *really* clean. But it requires deep bedding to trap the urine, and to keep the concrete warm in the winter. Wooden floors such as might be found in an old brooder house or similar building also require deep bedding. Moreover, they need frequent cleaning and disinfecting. This is the least desirable type of floor. Porous floors—earth, sand, or gravel—stay drier and therefore are easier to maintain.

Most goat pens are cleaned twice a year—spring and fall. It's amazing how well-tramped-down bedding keeps animals cleaner and produces less odor than bedding that is cleaned out weekly or more often. Of course, make sure the wind is not blowing toward the house at cleaning time! What you're actually doing is turning a well started compost pile, and the ammonia can be overpowering. And speaking of compost, some people use compost activators in their goat barns to speed up the process.

Bedding

Bedding is partly an individual choice and partly an economic one. Find out what's available in your area. Some people don't like sawdust. Others use it and like it. If it's free for the hauling when straw is $1 a bale can you blame them? Some people use shavings, or old hay, or chopped

This wooden manger is used for feeding hay and grain.

cornstalks, or ground cobs, whatever is available and works.

In some barns with less than ideal mangers, the goats pull enough hay from the mangers and spread it around enough so you seldom have to worry about bedding! This isn't the recommended method, and it isn't cheap. On the other hand, the organic homesteader can always take consolation in the fact that alfalfa makes terrific compost.

This manger is made of stock fencing (which is constructed of welded rods). It's easier than wood to maintain. The shape of the openings prevents a doe from pulling out mouthfulls of hay which is wasted.

The Manger

This brings us to the next problem—the manger. Goats are very wasteful. And in spite of some of the rumors about their dietary habits, they are extremely fussy about what they eat. Most of them refuse to touch spilled hay or grain.

It's instinct. The goat has more stomach wall area for its size than almost any other animal. Consequently, it's highly vulnerable to worms. Nature has protected the goat by teaching it to reach up, not down, for food.

The only practical way to keep hay waste to a minimum is through use of a keyhole manger. This is a manger which has a seven-inch-in-diameter hole for the goat's head, which is positioned high enough so she has to step on a toe board to squeeze through. The bottom of the hole should be three feet from the floor. A four-inch neck space enables her to slide down to get at the hay. The goat can eat comfortably, but in order to get out she has to step up again. Thus, she can't back out with a mouthful of hay, most of which would fall to the floor.

Watering

Along with the manger, the most important piece of equipment is the watering utensil. As with food, goats will not touch contaminated water. Yet they need great quantities not just to thrive, but to produce milk, which is mostly water. You want to induce them to drink as much as possible.

Hauling water can be a major factor in your chore time. The ideal is to have an automatic watering system, heated in winter, at such a height and location that droppings cannot contaminate it and birds cannot drink from it (and sometimes fall in and drown).

Instead of an automatic system, you can have a bucket of water for each animal. The buckets should be heated in winter, and they should be placed out of the way of droppings and birds. You can set up a more sophisticated watering system by building a stand outside of the pen and cutting a head opening in the pen wall so the goats have access to the water. Having the water outside the pen will make certain that the goats won't get manure in their water. They won't be able to tip the pail, either. What's more, you won't have to open a gate and go through a pack of affectionate animals to lug water for them.

The Salt Block

A salt block is a must. Some goatkeepers use mineral blocks, others use loose minerals in self feeders, and some use both. Mineral feeders made for hogs work well for goats and can be hung at a convenient height on a wall.

Fencing

Outside the barn comes the major investment: fencing. Goats are among the most difficult livestock to contain, not because they can't stand being penned, but because they know that out there are rose bushes, apple trees, pine trees, and other good things to eat—delicacies that would soon turn the homestead into a wasteland. You need good fencing.

The points to remember when planning a goat fence are that goats are excellent jumpers, they're very good at squeezing through small spaces, they like to stand on things, and they like to lean and rub on objects such as fence posts. The fence should be at least four feet high. Woven wire will soon be pulled down as the goats stand on the wire to look over unless some form of protection is provided. Posts should be stout and firmly planted so the goats don't push them over and simply walk over your fence. All this makes goat fencing expensive.

Perhaps the best fencing for large areas is tightly stretched woven wire with a strand of electric fencing running inside of it, two to three feet off the ground, to keep the animals away from the woven wire. Electric fencing alone does work, but not if a goat really and truly wants to get out. If a large area is to be fenced with charged wire, a smaller training area should be used to teach the animals the penalty for messing with the fence.

Large pasture areas require a lot of fencing, but many homesteaders and even commercial dairymen prefer to keep their animals in drylots. That is, they have a small exercise area, but all the feed is brought to them. Goats aren't good grazers to begin with. They're browsers, preferring to nibble here and there on bushes and trees. Most homesteaders simply don't have the room or the type of forage to permit goats to live on pasture. If there are trees or bushes, the goats will kill them. And commercial dairymen say they can't control the flavor of the milk if the goats are allowed to browse, since many wild plants impart particular tastes to milk. Small areas are much easier and cheaper to fence, and even when they have access to pasture, most goats will spend all their time laying in the barn, anyway!

The very best goat fence is stock fencing made of quarter-inch steel rods. Compared to some other fences, this is expensive, but since a goat can kill a couple of young fruit trees faster than you can run from the house to the orchard, good fencing can be cheap insurance.

Shade should be provided for hot summer days. Whether your goats are fenced in a drylot or a pasture, make sure there is a good shade tree or shelter they can get under.

FEEDING

Proper feeding is of paramount importance in any livestock operation. An animal needs good nutrition for good health just as you do. In addition, if you expect it to produce meat, milk, or eggs, it needs the extra raw materials that the production of these will require.

Moreover, feed is a major item of expense. The serious homesteader has to carefully consider feed to avoid paying downtown prices for his produce, after putting up the capital and labor.

The successful homesteader knows how to balance good health and production against cost. The young couple that couldn't find organically grown grain for their goat and enthusiastically fed her granola from a health food store, probably had a happy goat, but their milk must have been slightly more costly than vintage champagne.

Commercial Rations

Without a doubt, the easiest way to make reasonably certain your goats are getting the nutrition they need is by buying commercially prepared rations. In addition to the feed itself, you get the expertise of nutritionists, the benefit of least-cost formulation, with the mixing and grinding already done for you.

As Helen Walsh pointed out in *Starting Right With Milk Goats* (Garden Way Publishing) in 1947, a feed manufactured in her area contained yellow corn, crimped oats, bran, linseed meal, soybean meal, corn gluten meal, molasses, iodized salt, irradiated yeast, calcium carbonate, and dicalcium phosphate. Her observation: "What individual can attempt to assemble an assortment such as this for his goats? Yet these ingredients are, in the judgement of competent people, essential for a balanced diet."

Hay and Grains

Goats are ruminants, which means they require large quantities of roughage. Legume hays are desirable because of their high protein content. Protein is essential for growth, for the repair of tissue, for the development of unborn young, and for milk production. The homesteader, of course, gets much of his own protein from that milk. If only carbonaceous hays (nonleguminous, such as timothy) are available, the protein content of the grain ration should be increased.

The quality of any hay can vary considerably, depending on soil fertility, time of cutting, whether or not it was conditioned, curing conditions, shattering of the leaves, and length of storage. Always strive to get the best, for hay is the mainstay of your goats' diet.

Hay is generally fed free-choice to goats, that is, they always have it before them and can eat as much as they want. Grain, on the other hand, must be limited. A doe will gorge herself on grain and die. As a general rule, give each doe a minimum of one pound of a grain mixture a day. This is usually divided into two feedings. Milking animals get more: one additional pound of grain for each two pounds of milk produced. Some dairymen practice challenge feeding, which is gradually increasing the amount of grain until milk production reaches a peak, then decreasing the grain ration just a bit.

Mixing Your Own

If you are able to mix your own feed, here are a few suggested formulas from various sources which can serve as a starting point:

Formula #1
35 lbs. crimped corn
35 lbs. crimped oats
15 lbs. molasses
10 lbs. 30% protein supplement
2 lbs. salt

Formula #2
37 lbs. ground oats
35 lbs. gluten feed
25 lbs. wheat bran
6 lbs. yellow corn
20 lbs. molasses
12 lbs. cottonseed meal
9 lbs. soybean meal
2 lbs. salt

Formula #3
100 lbs. shelled corn
50 lbs. oats
40 lbs. wheat bran
35 lbs. soybean oil meal or cottonseed meal

Formula #4
100 lbs. barley
50 lbs. oats
20 lbs. wheat bran
20 lbs. soybean oil meal or cottonseed meal

Just as you can get 500 International Units of vitamin A from eating six stalks of asparagus or about half a tomato, so can you get the same amount of nutrients from a variety of feeds. Because of geographic factors, price, or anything else affecting availability, grains of similar nutritive value are always being interchanged. Substituting and interchanging grains can be a difficult, tedious process, especially if all factors of nutrition are taken into consideration (as they should be). The large milling companies have made the job easier for themselves by computerizing the chore.

If you want to substitute feeds, it will pay to do some work and find out just what's involved. Corn, for example, is just about equivalent to wheat for energy. But corn is the only cereal grain which contains appreciable amounts of vitamin A, while wheat is an excellent source of vitamin E. So corn is equal to wheat for feed value, but not in vitamin content.

Substituting sources of protein deserves special comment, because many homesteaders wonder how they can get around buying seed oil meal, the most common protein supplement. Oil meal is a by-product of oil extraction and, as such, is beyond the scope of the homestead. But feeding just soybeans (or linseed, cottonseed, peanuts or sunflower seeds) instead of the meal is no answer. Soybean oil meal is 45.8 percent protein. Soybeans are about 38 percent, but animals won't eat enough of them to significantly affect the total protein intake. If they're ground and mixed with other feed, the fat turns rancid. The oil meal is really the only alternative.

You can lessen the need for protein supplements by using the very finest legume hay possible, and by turning to higher protein grains. Corn, for example, has 8.5 to 8.9 percent protein. Wheat has more than 12 percent. But here again, if you don't use enough corn and have good enough hay (another source of vitamin A), you may have to add a commercial vitamin A.

The entire process of feed formulations is one of balancing one quality against another, and then introducing a cost factor. You can no more take a scoop of this and a handful of that and expect to come up with a balanced ration than you can take a little of this and a little of that and expect to create a cake.

Molasses is commonly incorporated into goat feed. Not only does it reduce the dustiness of feeds, especially ground feeds, it's also palatable and is a good source of iron and other trace minerals.

Feed for goats should not be ground fine, simply because the animals won't eat it if it is. But for greater digestibility and therefore feed efficiency, corn is cracked or crimped, and oats are frequently crimped or rolled.

Minerals are often added to complete feeds. Use any mineral mixture sold for dairy cows in your area, for mineral deficiencies vary geographically.

Once you settle on a feed, don't be constantly changing it. Corrections will be necessary as you go along due to price changes, because you will learn new things, or because something happens in your herd to let you know you were doing something wrong. But never change your basic feed formula whimsically. And don't toy with it without first determining the effect on total nutrition.

Forage

Organic gardeners like to experiment with their organic methods. Once a basic pattern is established, additions of kelp, cider vinegar, comfrey, and other organic favorites can play a role in developing your organic animal husbandry. By the same token, many goat enthusiasts take the time and trouble to forage for treats for their animals, especially those in drylots. Pine needles (rich in vitamins C and A), are relished. Alder, willow, apple, and other branches will be stripped of leaves and bark. Tree trimming always means treats for the goats.

While certain plants are poisonous to goats, this is a highly localized problem. Milkweed, wild cherry, locoweed, and others should be avoided. If in doubt, check with your county agent. The same plants that are poisonous to cows are poisonous to goats. Because of the goats' manner of taking a bite here and a nibble there in pasture, they seldom eat enough of a poisonous plant to cause harm.

While on the subject of pasture, I want to point out that a lawn is not a goat pasture. If staked out and starved, a goat will eat some grass, but not from choice. Don't expect her to produce milk on such a diet. In fact, in the spring when forage is lush and watery, a goat may eat a great deal of even good pasture and still be starving—her belly is full of water, not nutrients. Feed her well on hay before turning her out on lush grass to avoid bloat.

One other word of caution: mangel beets, which are easily grown on the homestead as stock feed, cause sterility in bucks.

BUCKS

"The sire is half the herd" is a common saying, and a true one. If you have half a dozen does producing 12 kids a year, each doe contributes genes to just two of them, or one-sixth. The buck, on the other hand, will affect *all* of them. Use of a poor buck will pull the quality of your herd down faster than anything else except general poor care.

The homesteader with a few goats probably won't have a buck. In the first place, it's the buck who gave goats their reputation for smelling. During the mating season especially, some of them get pretty high. This, along with some of their habits that delicate people find disgusting and repulsive, makes them a little less than the most welcome animal on the homestead. While most bucks are quite gentle and friendly, any male animal can be dangerous during the rutting season. A buck is a powerful animal, and one who was played with as a baby is apt to be dangerous without its

A) Toggenburg

B) LaMancha

C) Nubian

Bucks are more masculine—less "pretty"—than does.

owners realizing it until it's too late. And last but not least, a *good* buck (and any other kind should be in the deep freeze) is expensive, simply because they're the cream of the crop.

Even while there are disadvantages to keeping your own buck, the fact remains that he's a necessary part of the herd, if you want kids and milk. What's the alternative? The most common one is finding a doe in heat, putting her in the trunk of a car and driving off to a breeder who has a purebred buck. If she is actually in standing heat, that is, she'll allow the buck to mount, the service takes only a few minutes, even if you spend an hour or two driving. But that's still cheaper than the extra housing, feed, and labor involved in having your own buck.

If you have a number of does, it might be possible to lease a buck. In remote areas or goat-free regions, or even in locations with goats but none of your breed, it may be necessary to keep a buck. Buck kids are handled the same as does, but at four months of age or sooner they should be separated from the does. A four-month-old buck is capable of impregnating young does. Bucks grow more rapidly than does, but take longer to fully mature, usually three to four years. For this reason (and simply because they're larger) they require more feed. "The eye of the master" rules here: don't let a buck get fat, but feed him enough to permit him to grow to his genetic potential.

The separate buck facilities have one main requirement: they should be strong. I've seen sound 2 by 4's that were snapped like kindling wood by bucks. The ideal buck stall is probably made of iron pipe imbedded in concrete. Yard fencing could be concrete posts with iron pipe running through holes in the pre-cast posts. Wood, of course, is much more common. For fencing, the posts should be no more than eight feet apart, with 2 by 6's for the rails. Be sure to nail them from the inside so a buck pushing against the fence can't force the nails out. The fence should be at least four feet high.

For aesthetic reasons, the buck house is commonly kept at some distance from the rest of the homestead, but when planning its location, don't forget to consider ease and convenience at chore time. It's easy enough to neglect a buck when he's not being used when you can see him; tucked off in a corner somewhere, it's easier still. Make it easy to give him good care. Especially make it easy to clean the bachelor quarters, since good cleanings will reduce objectionable odors.

It is possible to deodorize bucks at the same time they're being disbudded. One of the main scent glands is in top of the head, and can be burned out with the disbudding iron. Better get help from someone experienced with this. Some people who take especially good care of their buck brush him regularly, trim his hooves at the same time they trim the rest of the herd, and even give him an occasional sponge bath with pine-scented deodorizing cleaner!

A buck is ready for service by the time he's seven months old. He shouldn't be used on more than a dozen or so does before he's a year old, but by the time he's two he should be able to serve four to five does a week.

Young bucks should not be pestered or played with roughly. This will make them unmanageable later, and a 200-pound buck goat who wants to "play" can be dangerous.

Artificial Insemination

Artificial insemination (A.I.) is just coming into its own in the goat world. Many breeders have tried it, without success, due to poor semen and improper techniques. In most cases it is not a substitute for a buck, as you still need one to determine whether the doe is in standing heat. But it is used, and it does work.

The obvious advantage of A.I. is that no matter where you live you can use the best bucks in the nation. An ampule of semen from a buck with a stud fee of $100 can often be had for $10. But—you'll need to make arrangements with an A.I. technician to use a $300-$400 liquid nitrogen storage tank for the frozen semen, and you'll have to acquire some equipment and skills. The homesteader with a few goats for milk for his own use isn't likely to find the results worth the trouble.

If goats grow on you, as they have a way of doing, artificial insemination is something to keep in mind as a good way to build a topnotch herd. Rather detailed information on A.I. is available from *Countryside & Small Stock Journal,* Waterloo, Wisconsin and from the American Dairy Goat Association, Spindale, North Carolina.

KIDS

It's always amazing to hear someone express surprise when they learn that you have to get a doe bred in order to get milk, but it happens regularly. Maybe this is just an indication of our attitude toward mammary glands—we've forgotten what they're really for. Nature produces mother's milk for the offspring, not for us. And a mother can't be a mother until she's bred. (There is such a thing as a precocious milker, one that lactates without being bred. Even very young kids can produce milk when this condition occurs, but it's relatively uncommon. For that matter, even some bucks produce "milk"!)

A doeling can be bred as early as four months of age, which is why young bucks and does should be separated. About seven months is the proper age. She still looks small then, but with proper nutrition she'll continue to grow as well as carry her young. Not only is there no advantage in waiting longer, but records show you'll actually get more milk from an animal bred early than from one who isn't bred until she's a year and a half old. Just consider the cost of the extra feed.

If you're concerned about doing things naturally, rest assured that in nature, where goats aren't penned at all, the herd sire will make sure none of his daughters will miss kidding by the time they're a year old.

Breeding

The first step in breeding is determining when the doe is in heat. Unlike cows, which can be bred the year around, goats have a limited breeding season, usually from September to January, depending upon weather factors. This limited breeding season is one of the chief problems of commercial goat dairymen. Since most does are bred in winter, when milk flow is normally less anyway (for cows as well as goats), maintaining an adequate supply of winter milk isn't easy. And demand is generally higher in the winter, too.

Knowing when a doe is in heat is important, and especially for the beginner with just a few animals, it can be difficult. If a buck is available, the job is simplified. The doe in estrus will climb the walls trying to get to him.

The buckless homestead can store up "canned buck aroma". Rub down the most odorous buck you can find with a cloth, or better yet, tie one around his neck for a few days. Put it in a glass jar, and close the lid tightly. Then, when you suspect you have a doe in heat, give her a whiff and watch her reaction.

There are other ways of determining if an animal is in estrus. A milking doe may drop in production or become a picky eater. She may be nervous, or bleat, or wag her tail more than usual, or have a slight mucous discharge. Any or all of these signs may be missing, and it's those does with "silent heat" that cause problems.

The average doe comes in heat every 21 days (unless she's bred) and will remain in heat from a few hours to three to four days. If you don't have a buck, you load her in the trunk of the car or even the back seat, and take her to the nearest purebred buck. If she's really ready she'll accept him immediately. One service is sufficient.

Stud fees range from nothing (with bucks that are worth about the same) to $100 for top animals. Expect to pay $5 to $10 most places.

If the doe doesn't come back in heat, you assume she's bred. There isn't much else to do at this point but to continue to give her good care.

The gestation period averages 155 days. If the doe is still milking two months before she's due to freshen, she should be dried off. Many goats will be dry by that time without any effort on your part. Others will dry off if you simply stop milking them and reduce their grain ration. The milk left in the udder will be absorbed back into the body. Still others are difficult to dry off. Just milk them as often as necessary, maybe once a day, then every other day, just so the udder doesn't get too distended.

The two month rest period is very important, as the young are developing rapidly at this point. This is a time to pay special attention to hoof trimming and to feeding. Protein content of the grain ration should now be about 12 percent. The doe should be in good condition, but not over-fat, as this can cause kidding difficulties. Minerals, especially calcium, are important for the dry doe.

Kidding

The birth itself will undoubtedly be one of the most exciting events to take place on your homestead. If you haven't had experience with this sort of thing and have the concern for your animals most goat owners seem to possess, the days before kidding will be anxious ones for you. Much of the literature dealing with goats discusses kidding problems. Descriptions of these problems scare the heck out of most people; it scared me when my first goat kidded.

I suspected my goat was ready because she had a stringy discharge, so I put her in a separate pen, well bedded with clean straw. I sat there and watched her for a few hours, but since she was quietly chewing her cud and I had to go to work in the morning, I finally went to bed. And the next morning there were two tiny, delightful kids, all dried off, standing on wobbly legs.

Of course, things *can* go wrong. In a normal delivery, labor can last for several hours or a fraction of that time. If it drags on and on and the doe appears tired, get help from a vet or an experienced stockman. Generally, though, you can rely on time and maternal instinct; the case where the doe needs your help is the exception rather than the rule.

In a normal delivery, the front feet and nose appear first. In a breech delivery (rear first) the kid will have to be turned around in order to be born. Wash your hands and arms thoroughly, preferably with a germicidal soap avail-

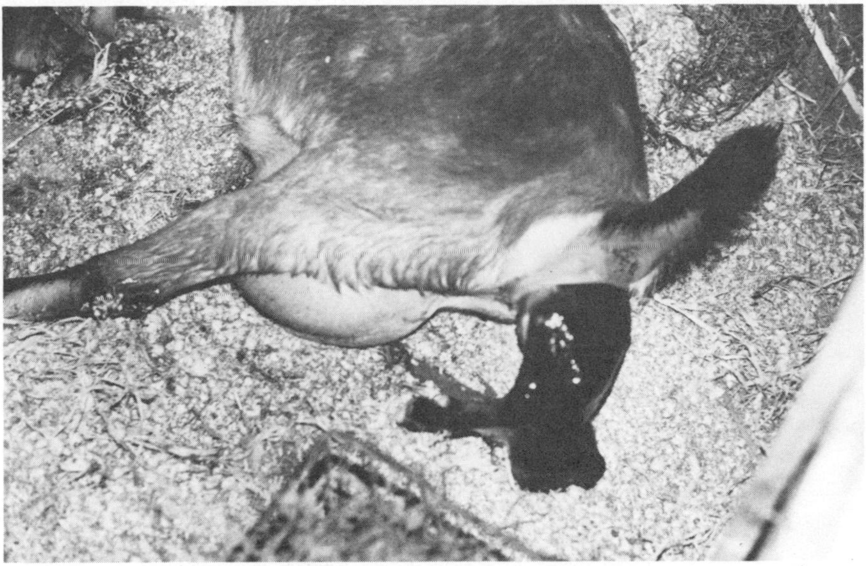

A goat kid being born. In most cases the mother needs little or no help, but you should wipe the kid's nose free of mucous and disinfect the navel.

able through veterinary supply houses. (These supply houses also supply antiseptic powders and lubricants.) You can then reach into the uterus and feel around to see what the problem is. Sometimes in mutliple births one or more of the kids becomes entangled in the cords, but a simple breech presentation is much more common. Simply turn the kid around so it can be born head and feet first.

Goats usually have two kids, three are not uncommon, and four or even five are not unheard of.

As soon as a kid is born, wipe the mucous from its face so it can breathe. Wait until the blood stops flowing in the umbilical cord. Then tie a string around it about four inches from the kids belly and another one about two inches farther away, and cut it between the two strings. Paint the cord with iodine to prevent infection.

Wipe the kids dry, and keep them out of drafts. Many goatkeepers place newborn kids under heat lamps, at least until they're thoroughly dry. Kid pens with solid walls, such as plywood, help prevent drafts.

After the kids are taken care of, the mother will appreciate a drink of warm or cold water, depending on the air temperature, and a treat.

Make sure the afterbirth is passed, then clean the stall. A retained afterbirth is a job for a vet.

Feeding the Kids

The kids can be allowed to nurse, or you can milk the dam and feed the kids. Either way, this first feeding is extremely important. The "first milk" or colostrum is a thick yellowish fluid containing antibodies critical to the kids' survival. The sooner you can get some into those little stomaches, the better. Some kids, however, just aren't interested in eating right away, and there isn't much you can do about that. Refrigerate the colostrum, then heat it later and try again. Use caution, or a doubler boiler: it scorches easily. In cases where there is no colostrum, for whatever reason, you can concoct a substitute with 3 cups of milk, 1 tablespoon of sugar, 1 beaten egg and 1 teaspoon of cod liver oil.

Most serious goatkeepers prefer to take the kids away from the dam at birth. If you want the milk, it's a necessity. Taking them away at once will eliminate many problems later. To do this, milk the doe and feed the colostrum to the kids. The oftener they can eat the first few days, the better

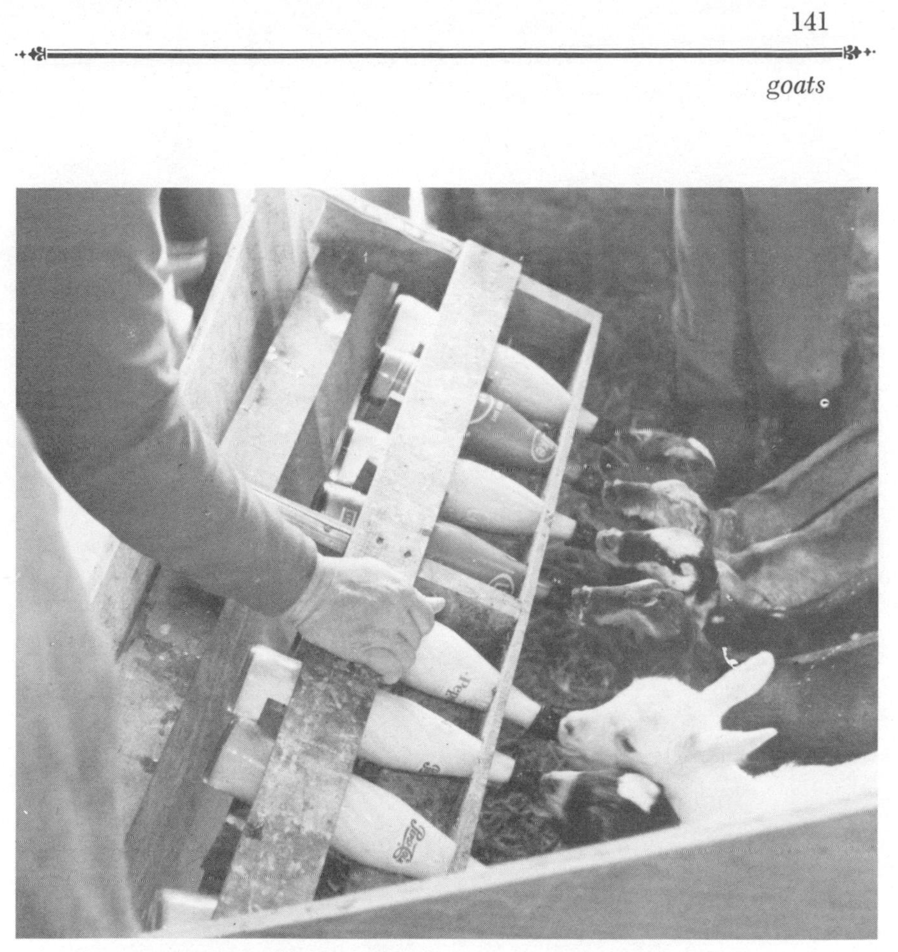

A bottle rack can save a great deal of time and frustration. The kids are quite violent with their nursing: the bar across the top is necessary to hold the bottles in.

they'll do. You can use bottles with lamb nipples or pans. Bottles are generally better; it's easier to teach a kid to drink from a bottle than from a pan. However, washing them for a bunch of kids can be a chore. Kids tend to drink faster from pans, and some people feel this tends to cause digestive upsets. You'll probably want to experiment with both systems and choose the one that suits you best. For larger herds,

A very nice, maintenance-free milking stand made from scrap metal.

there are special kid feeders that save a great deal of time and bother.

As a general rule, allow young kids four ounces of milk four times a day. Within a few days the same daily allowance can be divided into three feedings, if necessary. The milk should be about 100° F. The kids will always act hungry, but resist the urge to overfeed them. Overfeeding is one cause of scours, or diarrhea, which is the most common health problem in young stock. It can be fatal. If a kid develops scours, reduce the milk fed.

After the milk, give them as much warm water as they

will drink. That's a lot. Even very young goats will nibble at good, leafy hay. They'll start to eat grain, usually a calf starter ration, when about a week old.

Cleanliness is the number one rule in the kid pen. The bedding must be kept dry, and for kids this is a daily chore. Sanitary feeding utensils, it goes without saying, are absolutely necessary.

Weaning age is much discussed among goat breeders, and each one's opinion seems to vary with the amount of milk he produces and the disposition of that milk. If you have milk to burn, you'll certainly give it to the kids rather than dump it out. But if you're using or selling all the milk your does produce, you'll look for alternate ways of feeding those kids.

Milk replacers are used with varying success, dry skim milk is fine, and some people use cow milk. A kid should get whole milk for at least two weeks. If, by that time, she's starting to eat hay and grain, the amount of milk can be gradually reduced. Using good judgement and a good calf starter ration, a kid can be weaned as early as four weeks. Left with its mother, it may be five to six months before weaning.

GROOMING

The dairy goat herdsman must be familiar with several grooming techniques in order to maintain his animals in health and comfort. Disbudding, castrating, trimming, and hoof trimming fall into this category.

Disbudding

Disbudding is the technique used to prevent the growth of horns, as compared with dehorning, or the removal of horns themselves. Disbudding is fast, simple, and painless, which is more than you can say about dehorning. So disbudding is preferred. The job has to be done within the first few days after birth. At this time horn buttons are a fraction of an inch large and can be burned off.

Many people say they like horns and wonder why they should be removed. Even though horns may be beautiful and exhibitors are interested in beautiful animals, horns are a disqualification on show goats. This is because horns are just plain dangerous. Goats do fight, (there is always a boss goat in the herd), and if they have horns, they use them. Just as there are exceptions to every rule, there are people who leave horns on their animals and never have trouble.

The easiest way for most beginners to disbud kids is with caustic paste or sticks. It's usually easier to tell right after birth that the kid is going to be horned than a few days later. The hair is curly on the buttons of horned kids and straight on hornless ones. Clip the hair around this horn button with a pair of scissors. Then apply the caustic. If the kids wiggle or you have trouble getting the caustic in the right place, it would be a good idea to mask the area of the button with Vaseline. Cut a small circle of tape, about the size of the horn button, and fasten it to the button area. Spread Vaseline all around it. Remove the tape, and apply the caustic. This method is especially recommended for use with the paste type of caustic, which is somewhat more difficult to use than sticks.

The kid should be restrained for half an hour to prevent it from rubbing the caustic off and burning some other part of its body. Especially take care that the stuff doesn't get in its eyes. Some homesteaders merely hold the kid on their lap and watch TV for an hour after disbudding! Others keep their kids confined singly, in small pens, or cages. You'd never let a kid out in the rain anyway, but be especially careful to keep it out of rain after applying caustic. Rain would cause the caustic to run into the eyes and could cause blindness.

A much better way of disbudding is to use the hot iron. The iron can be the electric tool made for the purpose, or a rod of the proper size heated in a fire, branding-iron style. A good iron for kids has the diameter of a nickel. It should be hot enough to scorch wood.

The kid can be held on your lap in such a way that you can get a good grip on the muzzle with you left hand. The properly heated iron is then applied to the horn button with the right hand. If you have more than a few kids to disbud, consider making a kid-holding box. Even cardboard can be used. It must have a good slip-on cover. Cut a hole just large enough for a kid's neck in one end. The iron is applied for 15 seconds. Allow it to heat up again before doing the other button. You'll think you're killing the poor things, but just have a bottle of milk handy after the "operation" and see just how upset they really are!

Tattooing

The kid-holding box can also be used for tattooing, another job that will be necessary with registered goats. Tattooing is done with a special pliers containing letters and numerals made from small ink-covered needles. Goats are tattooed in the ear, except for the earless LaManchas which are tattooed in the web of the tail.

Some people ask if tattooing hurts. Judy Kapture answers this with a negative shake of her head, which makes the earrings dangling from her pierced ears jangle violently.

Castrating

Castrating isn't strictly necessary even for butcher kids. Most are slaughtered before breeding age, simply because it isn't economical to feed them out. However, bucks will have to be completely separate from doe kids, something that's pretty hard to guarantee with goats, no matter how good the fencing. Moreover, butcher bucks have a way of becoming pets, which means that, no matter how good your intentions, they become sires. Most serious goat raisers never let a buck bought as a pet or for meat leave their farm without being castrated.

The easiest way to learn to castrate is to have an experienced stockman or a vet do it for you, and watch carefully. It takes two people to do the job. One person holds the kid. He grasps it by the hind legs, with the kid's back against his chest. With a sharp, sterile knife, the other person simply cuts off the lower third of the scrotum and removes the testicles. Some people merely cut a slit in the side of the scrotum and remove the testicles through that. In either case, an antiseptic should be sprayed over the area after the operation.

There are several alternatives, which include elastrators, pinchers, and burdizzos. All crush the cords leading to the testicles, rendering them inoperative. Burdizzos are clamped on the cords and held for a few seconds. The elastrators (small, very strong rubber bands) work well and are easy to use, but they require a special tool that costs about $12.50.

Trimming

Trimming can be quite an art for show animals, but for homesteaders the goal is just to keep the milk clean. This means trimming the flanks and udder. Any clippers made for cows will work. Long-haired goats should be clipped over the entire body in the summer, not only to promote comfort, but also to discourage lice.

And finally, a goat needs its hooves trimmed several times a year. This is a simple thing, like trimming your nails, that is best done on a regular schedule just to make sure it isn't neglected. For a model hoof, look at a very young kid's.

When trimming a hoof, cut away all the hard outside edge. A good sharp jackknife works well, although some people prefer a pair of small pruning shears. Shave down the center part until the hoof is well shaped, like the hoof on that week-old kid. Stop cutting when the hoof starts to look pink.

MILKING YOUR GOAT

The purpose of keeping goats, of course, is milk. And just as with housing, equipment, and the animals themselves, there are a number of possible approaches to milking set-ups. The most basic system would be to squat on the ground and milk into a mixing bowl. Quaint, but backbreaking and not too sanitary. At the other extreme you could use milking machines with a pipeline system and a bulk cooling tank. Grade A, but hardly practical for the homesteader. (Regular milking machines can be adapted for goats.)

Equipment

The basic pieces of milking equipment are a milk strainer and filter disks, a small pail for the udder wash, storage containers, and a stainless steel pail with a half-moon cover. Another item which isn't absolutely necessary but is handy

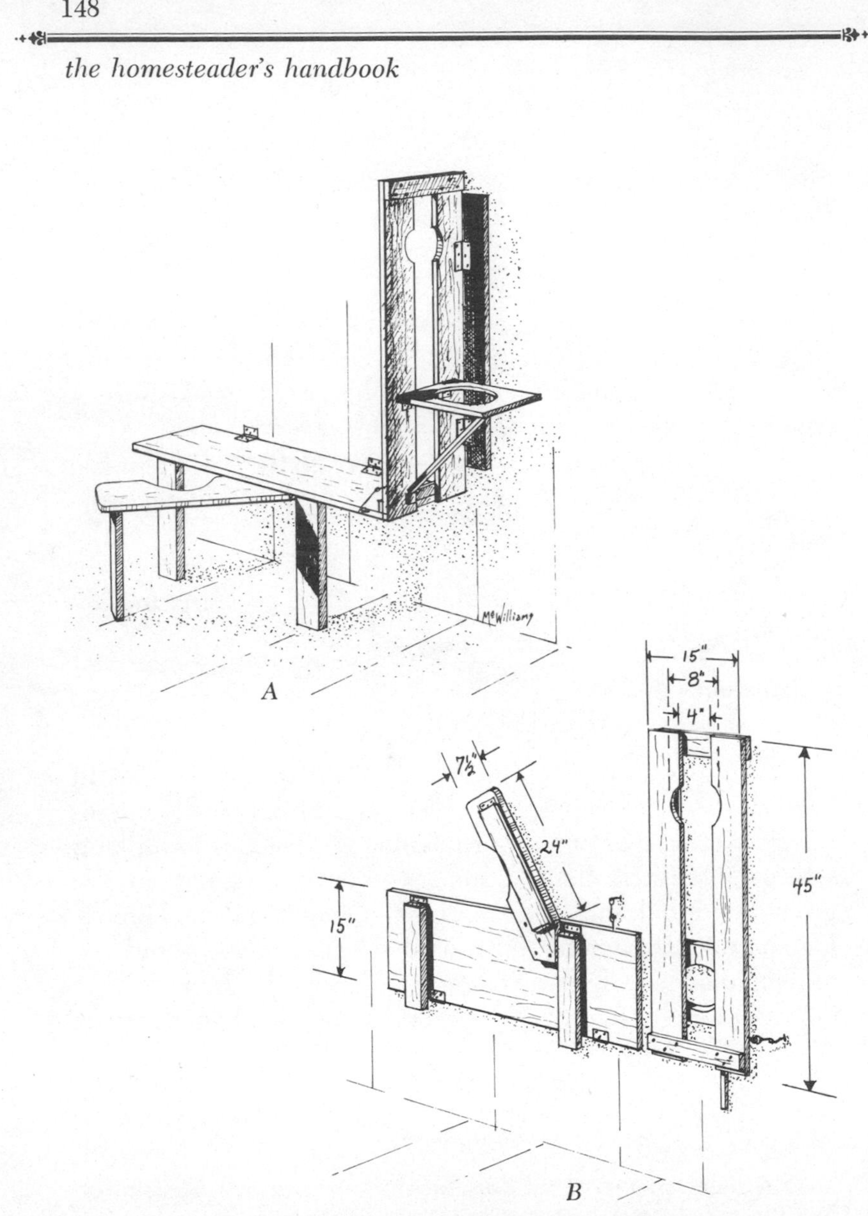

A) If you don't have much room in your "milking parlor" this home-made wooden milking stand may be right for you. B) It folds flat against the wall when not in use.

is a milk stand. (See illustration.) Some people get away without a strainer and strain through cloth; this is ridiculous.

Depending upon how much milk your herd produces, storage containers can range from fruit jars to two-or four-quart stainless steel or aluminum milk cans to 10-gallon milk cans. Smaller containers cool more rapidly, which is very important for quality milk.

Stainless steel pails with half-moon covers are expensive, but worth it. They last a lifetime, are easy to clean, easy to work with, and you'll want one eventually anyway. Some people use plastic pails and claim they don't have any problems with off-flavored milk or odors, but they don't say what the bacteria count of their milk is. If you simply must use something like this, you might want to take a tip from the lady who covers half of the top with plastic wrap to simulate the half moon cover of the real goat milking pail. Another idea is to cut a hole in the side of a gallon milk or bleach jug. This is easy to milk into and has a handle, but it is even harder to clean.

Keeping Clean

Speaking of cleaning, it's amazing that people who are down on commercial farmers and food processors for their use of chemicals can be so lackadaisical about home-produced food! For example, dairy farmers would lose their permits overnight if they cleaned their dairy equipment the way some goat milkers do. A handful of soap powder in the wash water and a glug of bleach in the rinse (if anything) won't do. Household detergents contain perfumes which leave a film on equipment and which may flavor the milk. Household bleach simply isn't pure enough for the dairy and contains compounds which may cause a medicinal taste in the milk.

Dairy cleaning agents fall into four categories: alkaline detergents, acid detergents, iodine compounds, and chlorine compounds. The detergents are for washing and the compounds are for sanitizing.

The alkaline detergent is the basic "soap" you use for washing equipment. Depending on how hard your water is, the alkaline detergent will leave behind a cloudy film which is called milkstone. It may not look like much to you, but to a bacterium it's home. Milkstone is often the cause of high bacteria counts in milk.

To get rid of the milkstone you use an acid detergent. You don't use it every day because it doesn't have the cleaning power of the alkaline detergent. Typically, dairy farmers scrub their equipment for six days with alkaline detergent. On the seventh day (when other people are resting) they scrub with acid detergent. Some farmers put acid detergent in the rinse water every day. This is probably a good practice for those who have hard water, because the minerals in hard water cause milkstone to develop faster.

Chlorine compounds do the best job of sanitizing, but they're too strong to use for washing udders. Iodine sanitizing compounds are used for udders because they're gentle and help heal small cuts and scratches. Some localities won't approve of them for washing equipment on grade A farms simply because they don't do a good enough job, but the trouble usually is that farmers don't let the equipment soak long enough. If you measure carefully and let the equipment sit at least five minutes, it will do the job, and you'll have one less bottle to clutter up your storage cabinet.

Be sure your measurements are exact when using any of these materials. If too weak they don't do the job; if too strong you're wasting money and risking contaminating your milk.

To properly clean equipment, rinse it with cold water immediately after use. If you have hot water handy, follow the cold rinse with a warm one. Never let milk dry in a pail or other piece of equipment. Wash the equipment with hot water and a proper dairy detergent. Scrub with a dairy brush, not a sponge or dishrag. Brushes are more sanitary; they scrape off films and get into crevices a cloth can't reach. Rinse immediately with hot water. *Drain* the equipment dry. Don't use a dishtowel!

Sanitize the equipment just before you use it again, with either the chlorine or iodine dairy solution. Submerge it for a few minutes, then let it drain for a couple of minutes. Don't rinse it off with tap water, as tap water is always full of bacteria.

There are obviously many home dairies that don't follow this procedure, which may be one reason goat milk has a reputation in some areas for tasting "bad." If one reason you produce your own food is to improve the quality, it doesn't make sense to cut corners. Commercial dairymen must follow these and other stringent measures to even be able to sell their product.

For a sizeable herd, the milking parlor is separate from the loafing pens. In the small home dairy, the milking stand is more frequently in the aisle. The barn, it goes without saying, must be especially clean and odorless then, because milk is highly sensitive to odors.

Milking

Ready to milk? Bring out the first doe, and once she gets the hang of the routine, she'll readily hop up on the stand, especially if you feed grain while milking. Most people do. It gives the doe something to do, and lets the herdsman keep a better check on how she eats. Any change in appetite is usually the first indication that something's wrong.

If the doe is especially dirty—a rare occurrence, and usually a reflection on the goatkeeper—she should be brushed. It's not a bad idea anyway. Should she shake herself off while you're milking, that clean wholesome food you moved to the country for takes a whale of a beating. The udder is washed with the sanitizing solution. Sponges made for cow dairymen work fine and can be boiled to disinfect them. Paper towels work well for drying.

The actual milking looks easy to the casual observer—until he or she tries it! The first time or two, expect to get more milk squirted up your sleeves than in the pail. The milker gets nervous, which makes the goat nervous, and the whole episode is likely to end in a draw when she either steps in the pail or simply lies down.

The process is simply one of trapping the milk in the teat by grasping it firmly, but gently, as close to the top as possible in the crotch of the thumb and first finger. Then, if you're doing it right, when you squeeze the rest of the teat the milk has no place to go but out. In slow motion, you squeeze the thumb and first finger together first, then press the second finger into the fist, then the third and finally the little finger. Release everything and start over again. It takes a little practice, especially to aim the stream of milk to go *in* the pail rather than *at* the pail, but some people catch on very quickly.

Some people are afraid of hurting the doe by squeezing too hard, an idea that should be quickly dispelled by watching a kid nurse or by letting one suck on your finger. Those little jaws and tongues are powerful. And, just as those kids butt and bump mama to make her give them more, the dairyman bumps or massages the udder as the flow stops to get more milk.

Stripping is the process of getting out the last drop by squeezing the top as in milking, but then instead of pressing in the other fingers, sliding the thumb and index finger down the length of the teat. Try to keep stripping to a minimum. Always remember that the mammary apparatus is not just a "bag." It's a highly developed, delicate mechanism. Milking requires a certain firmness, but it must be a gentle firmness.

The milk is weighed, usually to the nearest tenth of a pound, and recorded. This is very important even if you only have two or three animals. It's impossible to remember how much milk they all give, but this information will be of tremendous value when it's time to cull, to select replacements, or to sell an animal. You want to check the effects of changes in feed or other management practices on milk production. Going by memory, it's easy to say Susie is your best milker because you so vividly recall the day she gave a gallon. But it didn't last, and a doe with a less spectacular high day but a long lactation may actually be a far better producer.

After the milk is weighed, it is strained through a filter disk and cooled. Goat-size strainers are available from several of the goat supply houses. Cooling is perhaps the most important factor after general cleanliness in producing quality milk. The home refrigerator really isn't cold enough, although many people do use it successfully. The ideal is to place the jar or can of milk in a larger vessel of ice water. Milk is obviously at blood heat when it comes from the goat, or about 103° F. Some Grade A regulations require that milk be cooled to 38° F. within an hour.

Very little goat milk is pasteurized. Organic homesteaders are interested in raw milk because the heat of pasteurization destroys some of the nutrients. Furthermore it's a bother: another piece of equipment to purchase and clean. Moreover it's unnecessary. There has never been a documented case of tuberculosis in goats. Federal meat inspection figures for 1960 through 1965, and 1969 through 1971 show that more than 2,000,000 goats had been slaughtered under federal inspection, and not one carcass was retained or condemned because of tuberculosis.

Brucellosis, or Bang's disease, is no goat disease either. Judith Kapture, editor of the dairy goat section of *Countryside & Small Stock Journal*, obtained statistics on Bang's from the U.S. Department of Agriculture, then contacted officials in the states listed as having infected animals. Without exception, the figures from the federal government were false. In one state, it was "an office error." In others, the animals were retested and found to be negative.

One reason Ms. Kapture was particularly interested in these statistics was that some goatkeepers believe the only time they might get a suspect Bang's test on a goat is when she's pregnant or has recently freshened. It has something to do with the pregnancy, in other words, rather than with any disease.

If you keep your goats clean and healthy there's no need to heat-treat the milk. But if you insist, it should be heated to 143 degrees for 30 minutes.

CHEVON

In order to give milk, goats must be bred. Bred goats have kids. Over a period of time, half of these kids will be bucks. Most of the bucks will be worthless for breeding because they'll do nothing to improve the herd, and goats aren't monogamous anyway. Not all of the doe kids will be worth keeping either, and older animals that are no longer useful as milk producers will also be culled. The answer to this goat population explosion is chevon, or goat meat.

Chevon is very much like lamb, but because goats are dairy animals and are bred for different attributes than sheep—just as a Holstein has different conformation than a Hereford—it isn't economical to raise goats on the homestead for meat. The culls, on the other hand, must be disposed of anyway and can contribute much to the homestead larder.

One breeder in California has unlimited natural browse just below the snowline. Her goats feed on this all summer, with no hay necessary, and the young grow fat and sleek. At two months of age she starts giving the kids rolled corn and barley once a day. At four months they are butchered. "The flesh is well-marbled with fat like the best choice beef," she reports. "It is tender and juicy and fine flavored." She points out that a penned-up doe on alfalfa does not produce butcher kids like this.

Some homesteaders without the land or facilities, or without the time to devote to raising kids, butcher them at birth or as soon as they're dry enough to handle easily. These usually weigh about seven pounds. They are dressed out and cut up like fryer rabbits. The pieces are floured and seasoned and fried slowly for 15 to 20 minutes on each side.

Older cull does can also be used in the homestead kitchen. This meat can be canned in glass jars or used to make good jerky and salami.

Butchering

Equipment requirements for butchering are simple: a good sharp knife, about six to eight inches long; a larger, butcher knife; a stout rope; gambrel hooks which can be homemade or purchased; a meat saw; and an instrument for killing the animal.

Any animal to be slaughtered should be starved for 24 hours before butchering, but should be given plenty of water. Avoid exciting the animal in any way.

A carcass is just that: a hunk of meat. Getting it in that condition, especially the first few times, can be an emotional experience. Most people prefer to shoot the animal. The best way to do this is to tie it up and shoot it with a .22 at such an angle that the bullet enters the brain. If you hold the gun a few inches from the back of the head, the animal never sees it and is not frightened.

Lacking a gun, a sharp blow on the back of the head with a heavy hammer will do the job.

After the animal is dead, you should cut close behind the jaw to permit complete bleeding. Removing the head immediately makes the job easier for some people, since the animal no longer looks like a goat.

The Greeks, who have a great deal of experience in butchering goats, cut a small incision in the skin between the hind legs, and blow the hide (or fell) up like a balloon. This is said to separate the hide from the carcass and make skinning easier and cleaner. Some people who have tried it prefer to use a hand tire pump. Still others who use this technique insert a garden hose in the incision. The water is turned on until the hide is full, and when the hose is removed the fat usually seals the hole. The skin pulls away from the carcass as with air, but in addition, the cold water helps cool the meat.

You can skin by either laying the carcass on a table or by hanging it from hooks through the tendons of the hind legs. With a sharp knife (and have a stone and steel handy to keep it sharp), cut a slit down the belly, starting between the hind legs and going to the throat. Don't cut too deeply; you don't want to cut into the fell. Getting started is usually the hardest because once begun, you can work your fingers beneath the skin and hold it away from the body. At the ends of this cut, continue out along the insides of all four legs. Girdle the legs near the first joint, and the skin is ready to be removed. This is largely a process of pulling and peeling, with occasional help from the short knife to separate the hide from the fat and tissue.

Eventually you'll have a mantle of pelt attached only to the head, if it hasn't already been removed, and at the tail. Cut the skin off the head as close to the ears as possible. Then cut around the anus and loosen it until about of foot of colon can be pulled out of the body cavity. Tie the colon with a strong cord to avoid possible contamination, cut it off above the string and let it fall back into the body cavity. The skin can then be cut off at the base of the tail. Some people take care to save the tail, claiming that it makes an excellent stew.

The carcass is then opened by cutting down the belly from the cut made when tying off the colon, to the brisket. Be careful not to cut into the paunch. Let the paunch and intestines roll out and hang. Pull out the loosened colon end, work it down past the kidneys, and carefully remove the bladder. Tear the liver free and remove the gall bladder by cutting a piece of the liver off with it. If the gall bladder should break, wash the liver in cold water immediately. Then cut the gullet, and the offal will fall free. Split or saw the brisket, and remove the heart and lungs. Clean out any remaining pieces of tissue, wash the carcass with cold water, and wipe it dry.

If you haven't already done so, now remove the head.

Split the skull to get the brains (they can be used in sausage if they don't otherwise appeal to you), and cut out the tongue. Wash the liver, heart, and tongue in cold water, and hang them to chill.

As a last step, it's a good idea to check the stomach for worms and the lungs for signs of parasites or disease. This might well serve as an early warning system for problems that might be developing in your herd.

If you also raise pigs, clean, disease-free entrails can go to them. Otherwise bury them deep in the compost heap or in an unused section of the garden. Either way, nothing is wasted, and you're closing the cycle on a natural chain of events.

Aging

Aging is a process whereby the meat is tenderized through the natural action of enzymes and other ripening agents. The ideal temperature for aging is between 34 and 36° F. Even if this ideal can't be met, prevent freezing, or temperatures above 40° F. Freezing harms the quality of the meat, and temperatures above 40° F. permit mold and other surface contamination to develop. At the ideal temperature, a clean well-finished carcass with no nicks or gashes from skinning and a good covering of dry, unbroken fell or fat, will keep for two weeks or more without danger of spoiling. At higher temperatures (over 36° F.) the aging period should be shorter. Chops and roasts require more aging than stewing or braising cuts.

The aging process is often one of the main drawbacks to producing prime cuts of meat on the homestead. It's difficult to maintain the proper temperature without mechanical cooling. Aging, however, creates a difference that will be only of crucial importance to a real gourmet. If conditions are such that your meat can't be aged at all, there's no real loss. It's certainly edible, and probably still better than what you can find in the meat counter.

Cutting

Cutting the carcass—and doing a good job of it—probably requires more skill and experience than the actual butchering. The carcasses of goats and sheep are very similar, and the two are cut up in the same way. For a good diagram of how a goat or sheep carcass should be cut, turn to the chapter on sheep. These cuts are made with a large butcher knife and a saw, where necessary. (A good clean wood saw works fine.) A meat cleaver comes in handy for certain cuts. With the carcass cut into these basic pieces, it isn't too difficult to finish them up just by the way they look, according to your family's tastes.

Save trimmings and less desirable pieces of meat for sausage. They will have to be mixed with pork because they lack the fat necessary for good, juicy sausage. Wrap the finished cuts for the freezer, and the job is done. Chevon can be prepared according to any lamb recipes. Oregano is the favored herb for use with chevon.

MAKING GOAT MILK CHEESE

The average homesteader anxiously awaits the day when his goat herd produces enough milk so he can make cheese. Some of the best cheeses in the world are made from goat milk. France has a great many goats, and virtually all of their milk production is made into cheese. Cheese which brings exorbitant prices in the United States, it goes without saying.

Cheese was "invented", so it's said, when a man poured some goat milk into his goatskin canteen and took off on a journey across the desert. Some time later he sat down for a break, and took a drink. All that came out was a pale, watery liquid! Curious as to what happened to the milk he had put in there, he cut open the jug, and found a handful of a white mass of curd. He tasted it. . .and cheese was invented.

The process has been refined down through the ages so that now there are hundreds of different kinds of cheeses with their own distinct flavors and textures. The type of bacteria which sets the milk, the temperature of the milk, the amount of pressing, the length of curing, the curing temperature, the humidity, and many other factors affect the finished product.

The basic cheese recipe for hard cheese is as follows, according to a leaflet published by Charles Hansen's Laboratory, Inc., 9015 W. Maple St., Milwaukee, WI 53214, and included with the rennet tablets they sell:

Warm eight quarts of milk to 86 degrees. It doesn't have to be from one milking; it can be a week old, if properly refrigerated. Add one-quarter of a rennet tablet to a glass of cold water, and crush it with a spoon so it dissolves. Add the solution to the milk, stir, and leave it in a warm place, undisturbed, until a firm curd forms. This takes 30 to 45 minutes.

Test the firmness of the curd with your finger. Poke your finger into it and lift up. If the curd breaks clean over your finger, it's ready to cut.

The curd must be cut into small cubes about three-eighths of an inch square. To do this, cut it with a knife or spatula long enough to reach the bottom of your utensil. Hold the knife vertically for the first cutting, which will result in vertical slabs of curd three-eighths of an inch wide and as deep as your kettle. Then at right angles to the first cut, cut again, but this time hold the knife at an angle so that you are cutting diagonally into the depth of the curd. Then turn the kettle and make similar angled cuts from the other side.

With your hand stir the curd very gently, but thoroughly, for 15 minutes. Cut up larger pieces that come to the top. Then heat the curds very slowly, about 1½ degrees every five minutes, until the mass reaches 102 degrees. Stir with a spoon frequently enough to keep the curd from sticking together. When finished, the curd should hold its shape and readily fall apart when held in your hand (without squeezing) for a few seconds.

Remove the curds from the heat, and stir every 10 to 15 minutes so they don't mat together. In about an hour, the pieces, when pressed together in your hand, will easily shake apart.

The curd is then poured into a cheesecloth. Drain briefly, then place in a pail and sprinkle one tablespoon of salt over the curd. Mix it well with your hands. Don't squeeze the curds. Sprinkle in another tablespoon of salt and mix again. Then tie the four corners of the cheesecloth together and hang it over a kettle to drain off. This takes one-half to three-quarters of an hour.

After the curds have drained, take the ball of cheese from the cloth and place it on a table. Fold a long cloth into a bandage about three inches wide. Wrap it tightly around the ball, and pin it in place. Use your hands to press the top

A home-made cheese press made from a 3-pound coffee can and scrap lumber.

down. It should be smooth, or cracks will allow mold to penetrate to the center of your cheese. The round loaf of cheese should be about six inches across. If it is any larger, it will dry out too much while aging.

Place three or four thicknesses of cheesecloth on top of and under the cheese. Put the cheese on a board, place another board on top of the cheese, and weight it down with two bricks. The bricks are likely to tilt, and even fall off, as the cheese settles and compresses. The pressing job is easier if you can buy or build a cheese press. At night, turn the cheese over and add two more bricks.

The next morning, remove the cloths from the cheese, and place it on a board for half a day. Turn it occasionally. When the rind is completely dry, dip the cheese in parafin heated to 210 to 220 degrees, or paint it on with a pastry brush or paint brush. Store it in a clean, cool cellar or similar place. Turn it each day for several days, then two to three times a week. It should be ready to eat in three to four weeks.

The whey left over from cheesemaking can be fed to hogs or chickens. It should not be fed to goats, as it will cause scours. Whey can also be made into Ricotta cheese.

One of the best sources of cheese recipes (and other information pertaining to using goat products) is "Caprine Cookery," available at the time of printing for $1.50 from The Minnesota Dairy Goat Association, Algernon H. Johnson, Rt. 1 Box 144A, Litchfield, MN 55355. The dairy goat magazines and newsletters frequently print cheese recipes, too.

SHEEP

GETTING STARTED

One of the books that helped many people get "back to the land" in the 1940's—*Five Acres and Independence* by M. G. Kains—dismisses sheep with one paragraph saying "Sheep have no place on the small farm."

I disagree. Since sheep require no elaborate or expensive equipment or housing, very little care except during lambing, and can produce prime carcasses on little grain, many homesteaders could find sheep profitable for subsistence farming or even for a small cash income. Just from the standpoint of homestead meat, a sheep is much less expensive than a cow and is much easier for a beginner to care for and to learn about livestock with. They will produce meat in less than six months, compared with 12 to 18 months for beef, and they are certainly easier to butcher.

Then there is the wool crop which even many non-sheep owners would be delighted to have for home spinning.

Of course, all these reasons for raising lamb hold no

water if your family is not interested in lamb or mutton. Without their cooperation at the dinner table the entire sheep project, or any other, will be a disaster, no matter how good the prospects look on paper.

As with all other livestock, the homesteader would do well to begin his sheep flock by first finding a nearby breeder in whom he has confidence and then starting slowly.

Finding a breeder you can trust is always important because if you don't know anything about the animals you'll have to take the seller's word for their quality. Likewise, a good breeder wants to see you succeed and will be able to provide a great deal of advice and help.

As is also the case with any class of livestock, start out with the very best animals you can find and afford. Upkeep is the same for poor animals and good ones, but the returns and sheer pride of ownership is much greater with fine stock.

In general, look for heavy shearing ewes with pink skins and fleeces free from dark fibers. Yearlings are to be preferred, although a sheep isn't "old" until it's six or seven, and they can live as long as ten or twelve years. At any rate, make sure they have all their teeth. Broken mouthed sheep, those with teeth missing, are over the hill. With experience, it's possible to tell the age of a sheep by its teeth. The lamb has eight milk teeth in the lower jaw. By the time it's a year and a half old the center pair is replaced by permanent teeth which are much wider and longer. The second set of permanent teeth comes in in the second year, the third in the third, and the fourth in the fourth. The sheep is then said to have a full mouth.

After that the teeth begin to slant forward, and as the animal ages the teeth spread out, wear down or break off. When all are gone, it's a gummer.

Skinny ewes may indicate an infestation of worms, one of the most common plagues of sheep raising; avoid such animals. Even healthy appearing stock should be quarantined for several weeks before running them with your flock to avoid problems.

You want good size for the breed, good conformation, good condition. You learn what good sheep look like if you shop around enough, or better yet, pay close attention to the judging at a fair or other sheep show. Check for hard or lumpy udders or other udder defects. Lame or sore-footed sheep may indicate foot rot.

Female sheep are ewes. The males are rams. Castrated males are wethers. The young, of course, are lambs.

If you have more than a few ewes or if you don't have a neighbor whose ram you can use, you'll also need a ram. Once again as with all livestock, the male is half the herd: get a good one. The ewe passes on her characteristics, good and bad, to one or two lambs a year. The ram will affect every lamb in your flock. A purebred ram is usually a good investment if only because he will sire more uniform crops of lambs. There are good grades, and even good scrubs, but you're taking chances.

The breed to get is largely a matter of availability and personal preference. There are nine breeds of sheep with record associations in the United States, and some are naturally more popular than others. Suffolks, with their black faces, are the most common registered sheep with almost 37,000 in two record associations. Hampshires are second with just over 23,000, and Corriedales have about 9,000 registrations. Although less common, Dorsets, either horned or polled (naturally hornless), are quite popular among homesteaders. It's been said that Dorsets are the best mothers. We have some and have noticed that they have

Sheep are good animals with which to begin homesteading because they are so easy to care for. The black-faced sheep here is a Suffolk, and the white-faced ones are Dorsets.

completely different personalities than the Suffolks which we also raise. Dorset ewes weigh from 125 to 150 pounds, and rams weigh from 175 to 200 pounds.

Homestead sheep can be kept with goats, but should not be run with cattle or hogs because of the danger of injury, especially to pregnant ewes. Sheep and goats both require the same type of fencing: both are a dickens to keep confined. Good sheep fencing is designed to keep dogs and other predators out as well as to keep the sheep in. Sheep are defenseless, and a ewe torn to shreds by a dog is not a pretty sight.

Sheep, unlike goats, are close grazers and will do a good job of mowing your lawn. They'll damage some trees such as pines, but are nowhere near the problems goats are in this respect.

Like all animals, sheep require a constant supply of fresh, clean water, and salt. They need protection from hot sun—either trees or some form of shed—but they don't mind rain like goats do.

The most serious problem the small flock is likely to have is with worms. Several good wormers are on the market, but the best cure is prevention. This can be accomplished by rotating pastures so the sheep don't graze too closely, thus picking up the worms as they graze.

FEEDING

One of the attractions of sheep for the homesteader with waste land or pasture is the fact that the animals are grazers. Sheep on maintenance—that is, not producing milk, not being fattened for slaughter, and not pregnant—can live entirely on good pasture. The number per acre varies with the quality of the pasture, but you can figure about eight sheep will live on land that would support one cow. The best pasture can support as many as 15 ewes and their lambs per acre.

Remember to rotate pastures to prevent your sheep from picking up worms. Native grasses are acceptable, but timothy, sweet clover, ladino, alfalfa, brome, and orchardgrass are recommended.

On poor pasture or carbonaceous hay a grain supplement should be fed. (Carbonaceous hays are non-legumes—timothy, brome, etc.—and contain less protein.) One-third to one-half pound per day is sufficient. The grain ration can include ground or rolled shelled corn, oats, milo, or barley, depending upon local availability and price.

Without pasture, they'll need two to four pounds of hay and about a pound of grain consisting of 60 percent oats, 25 percent corn or sorghum grains, and 15 percent wheat bran. With grass hay, the grain ration should be 30 percent oats, 30 percent corn, 20 percent bran, and 20 percent linseed meal.

BREEDING

Sheep are seasonal breeders. Although there are breed differences and variations due to geographical factors, the usual period is from December through April. Heat periods occur every 13 to 19 days and range from three to 73 hours in length.

Before the start of the breeding season ewes should be eyed and tagged. Tagging is clipping wool from around the dock. Not only does this help keep the animal cleaner, especially when turned on green pasture which has a loosening effect on the bowels, but sometimes ewes fail to get bred because the wool or tags prevent the ram from properly connecting.

Some breeds of sheep are open faced, that is, wool does not grow around the eyes. The others must be eyed: the wool is clipped. Wool blindness can otherwise result. There is special danger from grass seeds and similar material collecting in the wool and permanently damaging the eyes.

This is also a good time to check the feet. Small farm flocks that don't do a lot of walking on rough terrain don't wear down their hooves like range sheep do. They can then accumulate filth and run the danger of foot infection. The hooves are trimmed just like goat hooves, with a sharp knife or pruning shears.

The average ram can service 30 or more ewes. Young rams should be limited, although a nine to 10 month old well-developed ram can breed 10 to 12 ewes. Rams can be used for about five years, although vigor and fertility are likely to decrease after that age.

There has been some controversy about the age at which to breed ewes, but research indicates that ewe lambs can be bred successfully without damaging them or their productive life span. The ewe should be at least nine months old. She'll require a better ration than the older ewes because she's still growing herself as well as producing a lamb, and she's somewhat more likely to need help at lambing time. But she'll produce more meat over her life span than a ewe that's a year old when first bred.

Sheepmen flush their ewes about two weeks before breeding. Flushing is simply putting the animals on lush pasture or if they aren't on pasture, feeding good legume hay free choice or increasing the grain ration to between one-half and three-fourths of a pound per head per day. They gain weight rapidly, shed more eggs, and multiple births are more likely to result. If you can have twins instead of single lambs, your meat supply will double without increasing the size of your flock, so it makes sense to try for multiple births. A 150 percent lamb crop is the goal to shoot for, that is, each ewe in the flock should average one and a half lambs.

Overfat ewes, on the other hand, are difficult to get bred. If they are too fat, feed should be limited at least six weeks before breeding starts. Then they can be flushed ten days to two weeks before the breeding season begins.

Proper feeding of the pregnant ewes is of extreme importance. During the first half of the gestation period the fetus is growing rather slowly and nourishment needs don't increase much. Overfeeding can be dangerous at this point.

But during the latter stages of the 143 to 151 gestation period the lamb is growing rapidly and the demands on the mother's body are much greater.

During the first half of the pregnancy good legume hay fed free choice will supply the nourishment needed. If only grass hay is available it should be complemented with one-tenth of a pound of protein supplement such as soybean, linseed, or cottonseed meal. A good method of feeding sheep during this period is to let them glean harvested grain fields. Cornfields, for example, will provide some grain that would otherwise be wasted, some roughage, and probably some green weeds.

Starting about the eleventh week, one-fourth of a pound of grain should be fed even with good roughage. By the sixteenth week this can be doubled.

Also during this period many sheepmen feed about a pint of molasses (or one-fourth to one-half pound of dry molasses) as a precaution against lambing paralysis. This condition is thought to result from the inability of the liver to transform body fats into sugar. While grain in the ration and plenty of exercise during the last six weeks of pregnancy are also good preventatives, the molasses still makes good feed. Lambing paralysis, also known as pregnancy toxemia or Ketosis, is characterized by listlessness, aimless walking, twitching muscle, grinding of the teeth, coma, and death. Overfat ewes who are underfed in late pregnancy are especially vulnerable, and the condition is more common among ewes carrying twins or triplets. In fact, it's often called twin-lambing disease. There is no cure, so prevention, through proper feeding, is the only solution.

Proper feeding of the pregnant ewe is important for the development of the young and to protect the health, milk flow, and productive life of the mother. Ewes in a weakened condition are more likely to disown their lambs than those in good condition. Undernourished ewes are more likely to have dead or crippled lambs. And yet, overfeeding is probably more of a problem among homesteaders because they generally really care *about* their animals as well as *for* them. You can kill animals with kindness.

A good ration during late pregnancy is a mixture of five parts oats, three parts shelled corn or grain sorghum, and one part each of bran, soybean, or linseed meal, and molasses. One-half to one pound per head per day should be fed with good legume hay. Salt should always be available free-choice.

The most critical period in a sheep's life is the first 48 hours. This is also the most critical period in the life of a sheep raiser! To be successful, you almost have to live with the animals at lambing time.

Losses can occur from chilling, dry ewes, ewes failing to claim their lambs, general weakness, and other factors.

Lambs are born like goats so experience with one will help with the other. The front feet and nose appear first in a normal delivery. In rare cases where labor is prolonged the mother may require help. Scrub your hands and arms thoroughly with a good disinfectant soap and investigate. As with goats, sheep are sometimes lodged in the birth canal backwards, sideways, or a leg can get tangled up and prevent normal delivery. By gently probing you can feel what's happening in there and take the necessary steps to expedite the birth.

When the lamb is born pinch off the umbilical cord four to five inches from the body and paint it with iodine. If the lamb gets chilled quick action is required to save it. Lambs that have appeared nearly lifeless have been revived by immersing the body in hot water up to the head for a few minutes. After all, it just came out of a sac of fluid that was 103° F. Dry it off and wrap it in a burlap bag along with a warm lamp, or in a warm place. Feeding warm milk as soon as possible will help too. Heat lamps are often used to help dry off and warm up new lambs. Pay close attention to fire precautions.

Another problem common with young lambs is "pinning." This is a condition where the first sticky feces glue the tail to the body and prevent further excretion. The tail should be loosened and the lamb cleaned up if this occurs.

Overfeeding the ewe the first ten days after lambing can be dangerous, as it increases the flow of milk too rapidly for the young lamb to consume, and udder problems can develop. In fact, with any livestock, it's best to hold the feed level steady for ten days after parturition to increase the milk supply slowly. Ewes on pasture won't need any grain until they're bred again.

Creep feeding lambs—providing grain in special feeders lambs can get into but the larger sheep cannot reach—is often used by commercial sheep raisers to produce greater weight gains more rapidly. It can be practical even for the homesteader with a few sheep.

Lambs should reach a weight of 90 to 100 pounds at five to six months of age. This should put 35 pounds of meat on the table. Slaughtering sheep is the same as slaughtering goats, which is covered in detail in the goat section of this book. After their useful productive life is over, older sheep are slaughtered as mutton.

With a 150 percent lamb crop and lambs dressing out at 35 pounds, eight or nine sheep will produce as much

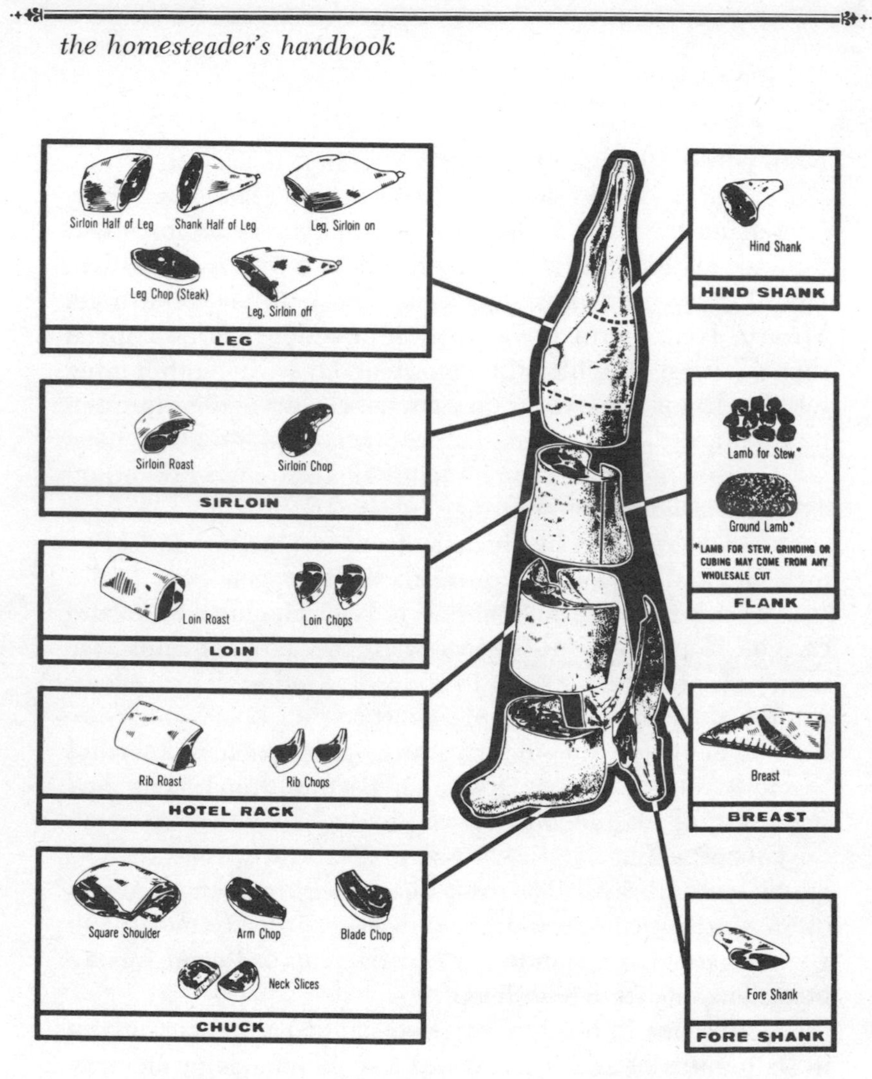

This lamb cutting diagram may also be used for goats, since both carcasses are so similiar. (Courtesy USDA)

meat as one cow. Moreover, they'll do it in six months compared with 18 for the cow. The advantages to the homesteader are obvious.

SHEARING

For ages, sheep have been as highly valued for their wool as for their meat. Synthetic fibers have changed that. However, the same nostalgia that urges homesteaders to get back to basics with home food production has also given rise to a surge of interest in spinning and weaving. Spinning wheels, long relegated to the strictly decorative antique department, are being manufactured again. Many areas have local spinning clubs where members help each other and experiment with such materials as even dog hair.

Your sheep must be sheared if only for their comfort so even if spinning doesn't appeal to you and you can't find a market you might be able to find hobby spinners who would be glad to eliminate the middleman and buy wool directly from you.

But first you must shear the sheep.

When it came time to shear my first sheep I asked around to see if any neighbors knew how. They didn't. One had sheep but he hired a shearer who had already been through the area.

None of the literature on sheep helped either. When they came to shearing most said it was extremely difficult to describe how to shear and the best way to learn was to work with an experienced shearer.

Every homesteader knows you can do anything if you have to. So I did it. The first one looked amazingly like my son after I had given him his first home haircut. Pretty shaggy. But it turned out that haircutting and sheep shearing were pretty similar, and both required a little practice.

There is no set time for shearing. The general rule is to wait until the weather has warmed enough to bring out the grease in the fleece (that makes cutting easier and the wool is said to be stronger) but before the animals go on pasture. Pasture has a loosening effect on the bowels, resulting in stained wool.

Before shearing, clip off any dung locks and grease tags (gobs of matted wool). Then start at the head, clipping as close to the body as possible. Work your way back to the tail, keeping the fleece in one piece. This is easier than it sounds because the wool fibers tend to cling together. To keep the wool as clean as possible, shear on a clean floor away from anything like straw or chaff, or spread out a large piece of canvas or plastic to work on. Avoid second cuts or trimming. Cut as closely as possible the first time. Second cuts result in shorter fibers that reduce the value of the wool.

I use a pair of hand clippers that cost me less than $3, although from my hair cutting experience I suspect that the $70 electric clippers make it easier to do a better job. With a small flock, the hand shears are entirely acceptable.

Check with your county agent to see if you have a wool market. In some areas the popularity of small farm flocks has resulted in pools where a number of farmers combine their wool to make a shipment. Special steps are required to prepare wool for market, such as tying the fleece with a special paper string. Your county agent can help with those details too.

Ten to 14 pounds of wool per ewe is a good goal to shoot for.

HOGS

HOGS ON THE HOMESTEAD

A few generations ago, no diversified farm worthy of the name was without a hog, at least to provide pork for home use. But as with most other aspects of farming, it's all but impossible to make a living raising hogs on a small scale today. So hog production has become highly specialized, and the average market hog is the result of antibiotics, concentrated feed, and mass production. As a result, home-produced pork will have flavor and tenderness you never thought possible. The average family will have enough pork by raising one hog at a time, twice a year.

Hogs don't require a great deal of room. When we lived in town, we always had one or two in a pen about 8-by-16 feet. The neighbor didn't complain. How could he when he had two sows with litters in his garage!

A properly built pen will help keep the pigs clean, and consequently keep odors to a minimum. Organic gardeners will regard the litter and manure as gold for the compost

Hogs are intelligent, affectionate animals that can grow up into pets—if you're not careful.

heap, so keeping the pen clean gives double satisfaction: caring for the animals and improving the garden.

Another point that might be of interest in passing is that some people actually enjoy keeping hogs. Small, slick piglets are obviously a joy, but even as they mature, especially when raised with plenty of personal attention, a pig can become a regular pet. Their intelligence is well-documented. They love to have their ears scratched, and if you regularly bring them warm whey or skim milk, which they love, they'll come running when they see you with that bucket, just as eagerly as any puppy. Just don't forget that their destiny is to end up on the dinner table.

FROM YESTERDAY TO NOW

Pigs were first domesticated in China about 4900 B.C., and Biblical writings mention them as early as 1500 B.C. From the beginning, pigs have suffered from a bad press, with "dirty pig" and "filthy as a pig-sty" and other similar phrases part of many languages. But these phrases are more a reflection on the farmer than on the animals.

Before the establishment of farms and settlements, hogs were not common livestock. They couldn't be moved by nomadic peoples, as could goats, sheep, and cattle. It is said that Columbus brought swine to America on his second voyage. There were only eight of them then, but 13 years later they had become so numerous that they were killing cattle, and settlers hunted them with dogs. In the 1630's swine were "innumerable" in the colonies. These hogs were described as black or sandy in color and weighing an average of 170 pounds.

The method of hog raising in those days will be of interest to homesteaders who dislike modern ways of producing meat animals. The hogs of early New England were allowed to roam free, unlike the cattle, sheep, and horses which were confined to the town commons. Nobody even "owned" them as individuals, because they were so numerous there was no such thing as a pig thief. (In the year 1790, six million pounds of pork and lard were exported from the United States.) At marketing time, the hogs were hunted down by dogs. A hound hung onto each ear until the animal could be tied and hoisted into a wagon. These semi-wild animals lived on roots, acorns, beechnuts, and such forage as they could glean from the forest. Hog husbandry was nil, except perhaps at farrowing time when a sow would be permitted to crawl under a barn for shelter.

Today swine rival rabbits in their variability under domestication and their amenity to human selection. In fact, the hog of today looks like a distant relative of those bred only a few years ago. It was relatively easy for hogmen to change from the chuffy, lard-type hog in demand at the turn of the century, to the long, meat-type hog producing the cuts desired by consumers today. However, like rabbits, pigs revert to their wild body form and characteristics in only a few generations. The hand of man is required in order to keep producing animals best suited for human food.

What are the differences between the modern and the wild hog? The modern hog has flesh on the sides and quarters, instead of being mostly bone with a hig head. The familiar razorback is a good example, for it's precisely the type of reversion we're talking about. The wild boar has an intestine nine times the length of his body, or a ratio of nine to one. The modern hog has a ratio of 13.5 to one. This makes the modern one a more efficient feeder. Also, a few paragraphs back, we mentioned that hogs in colonial days averaged 170 pounds. Today, mature boars of some breeds go well over 1000 pounds!

FURNISHINGS

Most homesteaders who are serious about raising good food at low cost will want to profit from the experience of the modern, specialized producers who *must* be efficient, then add or eliminate practices according to their own experience, demands, and situations.

The confinement system of hog raising has much to recommend it, especially on the small place where land is limited. The biggest problem for large producers is not the confinement itself, but overcrowding and the sheer weight

The logs placed against the fencing here prevent the pigs from working their way under the wire. This arrangement is fine for young hogs such as these, but bigger pigs would soon root their way under the fence.

of numbers, a situation the one-hog homesteader obviously need not worry about. Hogs can do well on pasture with just a simple A-frame shelter, but hog-tight fencing can be expensive, and pastures should be rotated because of the internal parasites hogs are subject to.

If you use fencing for pasture, erect it properly. With their tough snouts pigs will soon work under ordinary woven wire fencing. A strand of barbed or electric wire should be added to the inside of the fence about three inches off the ground to discourage rooting. Sows or boars that cause trouble by rooting are often ringed through the nose.

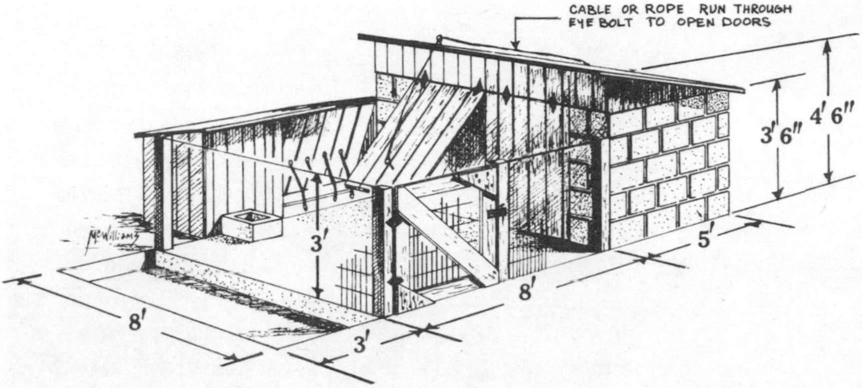

This is an ideal small hog set-up, with an easily cleaned concrete floor, a good stout shelter and enclosure, and easily accessible feeding and watering.

Shade is extremely important for hogs. Their inability to stand heat is one reason for the hog wallow, which is valuable for its cooling effect. In very hot weather it may be necessary to wet them down to prevent heat prostration. If some form of natural shade isn't available, as in a treeless pasture, erect a simple shelter, open on all four sides and as far off the ground as possible.

Almost any small building, or section of a larger building, can easily be remodeled to house a hog or two. As with any livestock, it should be dry and draft-free. As you design it, consider not only the ease and convenience you'll appreciate at daily chore-time, but also at pen-cleaning and butchering time. My first hog pen was at the back of our lot, a logical place for it, and an acceptable one when I did my own butchering. But when things got busy and I decided to send one hog to a slaughterhouse, there was no way to get the truck back to the pen! Something like building a boat in the basement.

GETTING YOUR HOG

The average homesteader would do well to forget about a sow. She'll produce seven or eight or 10 piglets twice a year, which puts you in the hog business when all you want is meat for your table. There are many additional problems with farrowing, not the least of which is maintaining a boar. It requires additional skills, equipment, space, expense, and labor.

It's much easier and more economical to buy a weaned feeder pig weighing about 40 pounds. This would be about eight weeks old. The hog market is highly volatile, and the price will fluctuate wildly. Just in the past few years it has varied from $11 to $40 in this part of southern Wisconsin.

The hog farmer should have wormed the pig, and he should also have clipped the needle teeth soon after birth. If a male, it should have been castrated too, changing it from a boar to a barrow. An old boar, castrated prior to butchering, is a stag. A young female, called a gilt, becomes a sow only after farrowing, or having a litter of piglets.

FEEDING

Young pigs up to 22 pounds require 22 percent protein in their feed for optimum growth and development. Up to 77 pounds it should be 16 percent, and finishing hogs can do with 13 percent. Your feeder pig, then, will need a diet containing about 16 percent protein.

A 40-pound pig will eat about 2.75 pounds of cereal grains a day and gain 1.10 pounds a day. Barrows will eat slightly more and gain slightly more then gilts, and consequently often cost a little more to purchase. Pigs this age will require about a gallon of water a day.

Pigs, unlike sheep, goats, and cattle, are single stomached animals. In fact, the pig's digestive tract is very similar to man's. This means pigs do not need the large amounts of roughage the ruminants require. Most market hogs in the United States are fed only corn, with a protein supplement. Partly because of greater feed efficiency, partly because of the necessity of mixing supplements into the grain, corn is generally ground. Hogs will eat corn even on the cob, but then you have the problem of getting them to balance the ration with the additional protein needed. Since corn is easily planted and harvested by hand, if necessary, it's a good choice of feed grain for the homesteader who can grow it.

But there are many acceptable substitutes. Wheat, in fact, has 105 percent of the feed value of corn for hogs, it is higher in protein, and produces excellent pork. Wheat should not be ground too fine (nor should any other grain), as it tends to become pasty and form balls of dough in the mouth, reducing palatability. The expense of what is usually a drawback, since it generally costs more than corn. And it's more difficult for homesteaders to grow using hand tools.

Barley has 95 percent of the feed value of corn; millet, 85-90 percent; sunflower seeds, 100 percent; and potatoes, 100 percent. However, because of palatability and the effect on the meat, many of these should not constitute 100 percent of the ration. Corn and wheat can be fed as a complete ration with protein supplements, but sunflower seeds and millet should be limited to 50 percent of the ration, and potatoes should not make up more than one-third of the diet. Oats have the same value as corn, but are too bulky for finishing hogs. Oats should be limited to one-third of the ration.

The homestead hog will also do well with alfalfa, comfrey, and Jerusalem artichokes, all favorite homestead crops.

The artichokes, in fact, can be dug by the hogs themselves, and if the patch isn't overgrazed they'll leave enough so you won't have to replant it. Being comparable to potatoes, artichokes should constitute only one-third of the ration.

Of course, none of these feeds have the required amount of protein, so supplements are required. Soybeans might come to mind first as a source of this protein, but it is not a good supplement for these animals. Not only are soybeans not palatable to hogs, but too much of them will cause "soft pork" which is flabby and oily and generally undesirable. Bacon from soft pork is well-nigh impossible to slice, even when chilled.

Milk is an excellent hog feed and a good source of protein. On many homesteads a pig can prevent a surplus of milk from going to waste. Moreover, a pig will relish skim milk for which there may be no other use, as well as whey which is left over from cheesemaking.

Although nobody seems to know why, it's generally agreed that milk and milk by-products help control some of the internal parasites of swine. Commercial farmers usually can't afford this organic method of parasite control, but homesteaders can. Not only do milk-fed hogs make faster gains, but the meat is excellent, as evidenced by the reputation of Danish pork which is milk-fed and exported as a delicacy.

Fifteen pounds of milk can replace one pound of tankage or meat meal which the homesteader would otherwise have to purchase. A pig can have one to one and a half gallons a day (eight-12 pounds). Since protein requirements decrease with age, this amount of milk will suffice in spite of the greater size of older hogs.

Peas are often used as a protein supplement in the North and Northwest. Peanuts are used in the South, but peanuts are like soybeans—too much of them causes soft pork.

Your homestead pig will eat about 600 pounds of feed from weaning to slaughter. If you have to purchase feed, find the local price, add the cost of 600 pounds to the original cost of your feeder pig, and you'll see how much your pig will cost at butchering time. In many cases, purchasing feed will make the homestead hog project unprofitable from a monetary standpoint, but chances are the end result will be worth it. You'll have bacon, hams, roasts, and chops you couldn't buy at the supermarket at *any* price.

Remember, too, that the pig will be your organic garbage can. Extra eggs you don't want to bother selling are an excellent source of protein. All clean kitchen and garden waste can go to the hogs, as well as offal from butchering chickens and rabbits. Pigs are carnivorous and eat meat, bones, and even feathers. Commercial feeds contain these ingredients in processed form.

Another relatively inexpensive way of feeding hogs, if the space is available, is to pasture them. Many seed companies sell special hog pasture mixtures that contain alfalfa, rape, field peas, soybeans, and other crops, depending on location. One acre of alfalfa or rape (which, although not a legume, equals alfalfa in feed value to hogs) can support as many as 20 to 25 hogs. For the homesteader with a confinement system, it can be a simple matter to cut the forage and bring it to the hog daily.

KEEPING THEM HEALTHY

About one out of four pigs dies before weaning age, a good reason why you should think twice before getting into a farrowing operation. But, another 2 percent die between weaning and marketing. Since this represents a tremendous economic loss, farmers quite understandably place great reliance on antibiotics. The list of swine diseases seems almost endless, and most of them are beyond the scope of a discussion here. However, there are four main reasons why most homesteaders have few problems with the health of their stock.

First, they can easily start out with healthy stock. It's obviously easier to find one healthy pig than a carload of healthy pigs. This is of prime importance, since most problems can be avoided by the simple expediency of starting out with healthy animals. It may take some experience to notice some slight problem, but in general, even the uninitiated can tell the difference between a sickly animal and a healthy one. You wouldn't pick a runty, listless animal or one with obvious defects, such as rupture or limp, but rather one with bright eyes, an alert nature, and a good appetite. Often the premises and manner of the seller will tell you a great deal about his stock.

To keep the animals healthy once they reach your homestead, take the measures necessary for proper sanitation. Frequent and thorough cleaning will help you more than anything else to raise your livestock without antibiotics and medication. This is one reason why concrete-floor-confinement housing works so well for the small place: it's easy to get really clean and to keep clean. In any event, never keep hogs in buildings previously used for the purpose without complete and thorough cleaning and disinfecting. Keep the bedding dry. Remember that sunshine and fresh air are the cheapest and best disinfectants. Rotate pastures, and don't fertilize hog pastures with hog manure.

The third health factor which often causes problems for the large farmer, but seldom for the homesteader, is overcrowding. With any livestock, not just hogs, you're more likely to have problems with large numbers than with just one or two, and not only because of the mathematics involved. When your entire herd consists of one animal, it's obviously easier to give it more attention, to notice when something is wrong, to keep it more comfortable. And one sickly animal in the larger herd can quickly infect many others that would have been perfectly safe on the one-hog homestead.

Good nutrition is also obviously essential for health, but here again, the organic homesteader can do a better job than most farmers with the same amount of knowledge, because the homesteader is producing quality, not quantity. Note, however, that the homesteader must know at least as much about nutrition as the farmer! The homesteader who doesn't use commercial supplements must be even more aware of the nutritional needs of his stock and of the nutrition in the feeds he provides. Minerals are important for hogs. If you raise them on concrete, provide a few shovelsful of uncontaminated sod, which contains trace elements needed by the hog.

BUTCHERING

At about six months of age, your hog should weigh close to 225 pounds. This is the time to butcher it. Additional feed will produce fat, not meat, and the rate of gain will be much less, meaning it will take more feed to produce each additional pound of hog. As is usual with slaughter animals, withhold feed (but not water) for 24 hours before butchering.

If you intend to have someone else do the job, consider how you're going to move the animal. You'll need a ramp to get it on a truck. Hogs are not always cooperative about being loaded. The best way to get your hog onto a truck is to have several people with gates, panels, or something similar press in on the animal until it has no choice but to go where you want it. It's also been said that when the Lord made pigs, he put the head on the wrong end. Put a bushel basket or box over the pig's head, then guide it *backwards* where you want it to go.

Butchering a hog is a more awesome task than butchering a chicken, a rabbit, or even a goat or sheep. But prior experience with small animals is invaluable. All of them have hearts, livers, stomachs, and intestines, and once you get involved in the job, there really isn't a great deal of difference between a rabbit and a goat . . . or a hog.

The chief difference is in getting started. Rabbits are pelted with the skin in one piece, or cased. Goats, sheep, and cattle are skinned by slitting the hide down the belly. Hogs are most commonly not skinned at all. They are scraped. You'll need a drum or tank or even an old bathtub that will hold the 200-pound-plus carcass, and a means of handling that weight: either a winch of some sort, or plenty of help. And you'll need enough water to cover that carcass, as well as a means of heating the water to 150° F.

I scald my hogs in a 55-gallon drum which has one of its ends removed. The drum is set at an incline at the end of a sturdy table. A good fire (corncobs burn with a vengeance) is lit under the drum, which is about half full of water. The hog, which has been dispatched with a rifle and whose jugular vein has been severed, is hoisted on the table. This is a two-man job. It is then dunked into the water and swished around a bit. It must be kept moving to avoid scalding.

It is then pulled out of the hot water onto the table, and scraped with bell scrapers (available from NASCO, Ft. Atkinson, WI) The process is repeated as often as necessary to remove the hair. Keep the water hot. A sharp knife can be used to shave difficult places, but this results in leaving whiskers not unlike a 5 o'clock shadow. Scraping removes all the hair. Even black hogs will turn white during the operation.

When the scraping is complete, carefully slit the belly as you would for a sheep or goat, remove the entrails, and wash the carcass with cold water.

Hogs *can* be skinned, and some people profess to prefer to handle them that way, but it's a more difficult job than skinning anything else because of the fat. You'll also lose a great deal of fat for lard-rendering and soap-making. Skinned hams and other cured cuts, won't keep as well as those which have not been skinned.

Whether scraped or skinned, the viscera are removed, the head is removed, and the carcass is split into two halves by cutting and sawing down the spine. You have two sides of pork.

The meat should be cooled overnight before further cutting, and since homesteaders don't have walk-in coolers, this means choosing the right time of the year to butcher. The meat shouldn't be allowed to freeze, but it should be cooled to 40° F. or less if possible. You obviously don't want flies or other pests around. This is one reason Indian Summer is the ideal time to butcher. But you have to plan ahead, and get your feeder pig at the right time.

Good hogs commonly dress out at close to 70 percent. That is, of the live weight of the animal, 70 percent is usable, 30 percent is offal and other waste. This 70 percent includes such delicacies as ears, brains, and tongue. Even the tail can be used to grease frying pans.

A chart of pork cuts is handy to have before you when

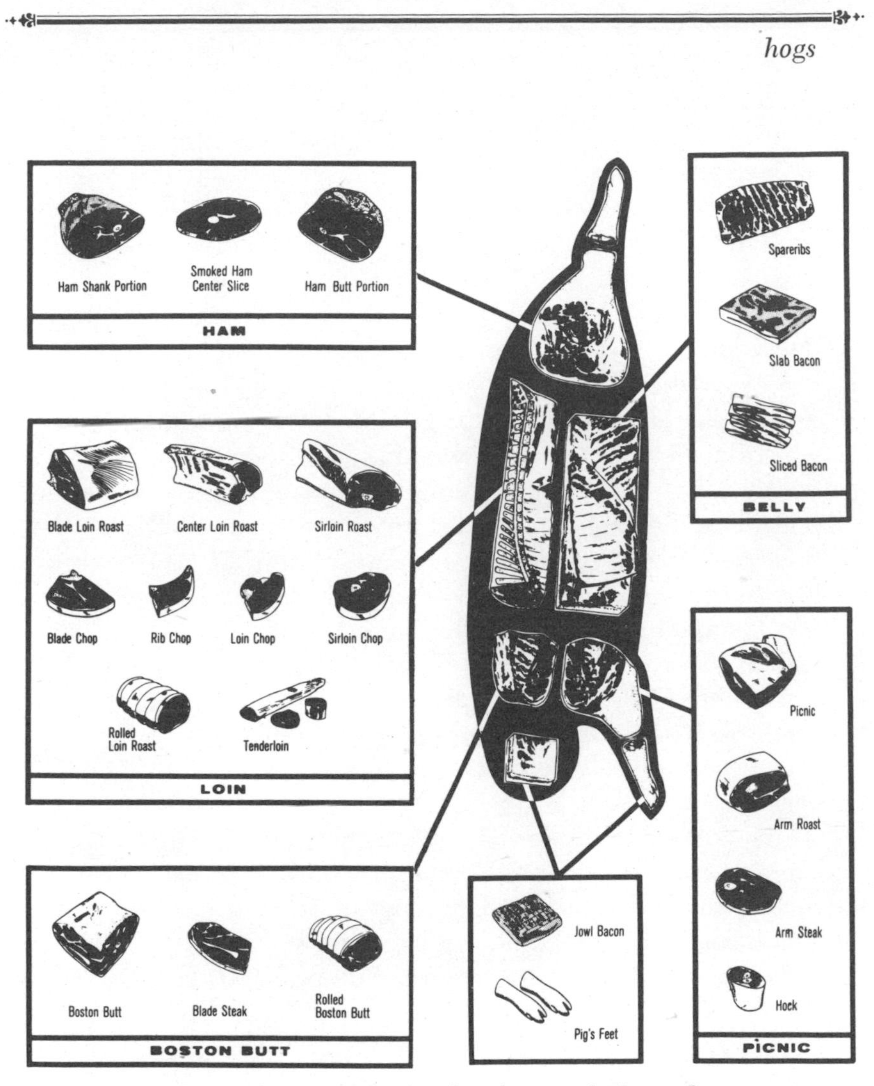

Butchering hogs is no easy job. It's a good idea to have a diagram like this one here when cutting up the carcass. (Courtesy USDA)

you start to turn those sides into chops, roasts, bacons, hams, sausage, and lard. A meat saw is a must, along with a good butcher knife. A meat cleaver comes in handy.

APPENDICES

Appendix A

HOW MUCH IS ENOUGH?

Along with the question "how?", most beginning homesteaders want to know "how much?" How many goats will it take to keep the family in milk, how many chickens for eggs, how large should the garden be?

Only you can answer these questions, and only through your own experience. There are many variables in eating habits, and even more variation in the production abilities of animals and land. We can talk about averages, but averages don't mean much if you're in the habit of eating every day.

Home food production is often a "feast or famine" proposition. One day you have eggs and milk in abundance, and another day your flock and herd don't produce enough to feed a starving mouse. It takes experience and planning to level out these peaks and valleys as much as possible.

Of course you need *some* idea of what to expect, so let's look at the averages. . .and some of the pitfalls that lie in wait for anyone who relies too heavily on them.

Let's start with an easy one. Chickens. You know, or can make a good guess, at how many chickens you go through in a year for frying, roasting, and barbequing. However, even this basic pattern is likely to be altered on the homestead. At first these home-grown birds will be so delicious you'll think you could eat chicken every day. But if you actually start eating chicken every day, you'll soon think you could never eat chicken again!

Most families fall into pretty set patterns. One that really enjoys raising chickens, has the proper facilities, a good source of feed or range, and doesn't mind butchering chickens, will consume a lot more than the family that has trouble keeping them out of the garden, has to buy high-priced commercial feed, can't stand butchering the things, and doesn't care much for chicken in the first place.

If you butcher them at light weights, you'll obviously need more birds than if you grow them out as roasters. But feed conversion efficiency suffers at heavier weights, and the meat will likely cost more. But as an average, figure on about 13 chickens per person per year.

Eggs? The average hen should lay about 220 eggs a year. Different breeds, and different strains of the same breed, can vary from that tremendously—generally a downward variation. Birds lay fewer eggs their second year than their first. In addition, environmental factors have a decided effect on laying. Feed can have an effect, especially if it is one of the modern, high-energy commercial feeds that help birds to produce more eggs. (Of course, they may be more expensive eggs too, than those produced on cheaper, homegrown grains and range.) Disturbing influences such as wild animals or even domestic pets can take their toll.

In the North especially, winter production is commonly suppressed because of colder temperatures which require more feed just to maintain body functions, but more importantly because of the decreasing amount of daylight. While a dozen birds may provide all the eggs you need during the summer, you may end up with eggless breakfasts in winter.

Most homesteaders who are serious about substinence farming want enough at all times. That means during the feast periods there is a surplus. These can often be sold (when prices are lowest because everybody else has a surplus too!) or they can be recycled on the homestead. Eggs, being excellent sources of protein, can be fed to any other animals with good results, especially young animals. Mix eggs thoroughly with the milk for kid goats or lambs or the calf. Pigs of any age will relish them. Eggs can be boiled and mashed and fed to chicks and even the hens themselves will eat them. (Some breeders of purebred cows and goats who are interested in production records rather than marketing milk often feed the milk back to the animal that produced it. Production increases, but to the homesteader, it's a vicious circle.)

Eggs can also be preserved in various ways for use in periods of short supply. The old standby used by the old homesteaders was to preserve them in waterglass. For shorter periods, and in cool to cold weather, you can simply pack eggs small end down in containers of sawdust or oatmeal.

Most modern homemakers find it more convenient to freeze excess eggs. The method of freezing depends on the end use of the product. For scrambled eggs and recipes calling for whole eggs, first decide how many you'll need, and freeze them in containers holding that amount. Mix them lightly, but don't beat in any air. Mix in either one-half to one-quarter teaspoon of salt per cup (six eggs) or a tablespoon of sugar, corn syrup, or honey. The choice of stabilizer depends on the end use, whether it's to be salty or sweet. Thaw them in the refrigerator, and use them within 12 hours.

For longer storage periods—up to eight months—whites should be separated from the yolks. The whites can be frozen in the quantities in which they'll be used. Nothing has to be added to the whites.

The yolks, on the other hand, will coagulate unless a stabilizer is added. They need salt, one teaspoon or so to six yolks, or sugar, honey, or corn syrup, two tablespoons for six yolks. Don't forget to label them if you make batches with different stabilizers.

One tablespoon of the yolks and two tablespoons of the whites reconstitute one whole egg.

Still another possibility is using the fresh eggs in baked goods which can be kept in the freezer until there are no eggs to bake with.

Milk is another feast-famine homestead product. The average goat—taking herd averages from established breeders with experience who have culled out the poorest animals—runs around 1500 pounds a year, or roughly 750 quarts. (Average per capita consumption of milk, in cheese, ice cream, and butter as well as in the fluid form, is about 570 pounds.) But even if you have an average goat, you can't count on having two quarts of milk a day throughout the year.

One basic reason is that the goat is milked for ten months, and then is allowed to rest for two months before kidding. What's worse, lactation is not steady. Milk flow increases gradually for about a month after kidding, then more slowly decreases. Many goats will be dry long before those ten months are past. You'll still be doing chores, buying feed and hauling manure, but you won't have any milk.

So once again, the serious homesteader who requires a year-round food supply will insure himself by having too much rather than too little. As with eggs, excess milk can be put to good use. Cheese making is a golden opportunity for the goat farmer with too much milk. Properly made cheese, with no cracks to permit mold to enter the interior, properly cured and stored at the proper temperature, will be at its best long after the milk flow has dwindled to a trickle. Ice cream and butter will keep for months in the freezer, and the milk itself can even be frozen. It tends to be somewhat watery when thawed out, so many people prefer to mix it half and half with fresh milk. If you have only half enough milk during the famine period of the cycle, this is the answer.

Milk, also, is excellent livestock feed and can be fed to chickens, hogs, calves, and even rabbits. Strict sanitation is important when feeding milk, as it quickly draws flies. Sour milk can be fed but then it should always be sour: don't feed sour milk one day and sweet the next or digestive upsets may occur.

The homesteader who is still in the armchair stage, planning, can easily conclude from the averages that two goats freshening at different times will take care of his needs. It doesn't always work out that way in practice. Better decide that anything that can go wrong, will. It may not be possible to get your animals bred when you want them bred, they may not be the good producers you had hoped they were or were led to believe they were, they may have short lactations, and accidents can happen.

All these problems can be overcome with time and experience. You have to get to know your animals, you have to be aware of your family's eating habits which will likely change when different foods are available. Then, with intelligent culling and planning, your homestead can become self-sufficient. Until then it's largely a matter of luck.

To look at the other side of the coin, it's possible to have too much too, of course, especially if there are no opportunities for selling surpluses and no way to recycle them through other animals, thus holding down feed costs. If carried to extremes, such overstocking can make your home-grown products very expensive indeed. We have to take a middle road.

How? Back to the averages. They are interesting and somewhat helpful not only for the planning homesteader, but also for those with more experience who want to see how they compare with the "norm."

Food item	*Average per person consumption per year*		*Average livestock units required for this production*	
Milk	*75*	*gallons*	*.50*	*goat (1500 pounds/year)*
Butter	*26*	*pounds*	*.33*	*goat (1500 pounds of milk/year)*
Eggs	*30*	*dozen*	*1.7*	*hens (220 eggs/year)*
Beef	*113*	*pounds*	*0.22*	*steer (1000 pounds liveweight or 465 pounds retail cuts)*
Pork	*72*	*pounds*	*0.72*	*hog (210 pounds liveweight or 135 pounds retail cuts)*
Chicken	*40*	*pounds*	*16*	*chickens, 2½ pounds dressed*
Lamb and mutton	*3.2*	*pounds*		
Veal	*2.3*	*pounds*		
Rabbit	*0.12*	*pounds*		

These are national averages from the U.S. Department of Agriculture and other sources. This accounts for the poor showing of rabbit meat, for example. If the homesteader raises rabbits, his consumption of other meat will decrease.

Butchered cull goats will alter the list. Vegetarian inclinations will change it even more drastically.

Each homestead is different and each varies from year to year as the family gains in experience, has good luck or bad, and tries new projects. The planning, with averages, is interesting and helpful, but the only really sound planning results from personal experience. This means, once again, starting out slowly, and growing into it.

Appendix B

MINERALS

Examining the labels on commercially prepared feeds will generally reveal that minerals have been added. The organic homesteader who uses home-grown feeds and is highly aware of the importance of minerals, is often concerned that something may be missing in the rations he provides his livestock. Acute mineral deficiencies are rare, for most of the 15 essential mineral requirements are provided naturally in feeds. However, some of the more important minerals, especially those required in special abundance by certain classes of livestock, may not be present in sufficient quantities in the feed because of soil conditions. Since organic farmers pay special attention to these trace elements, feed grown on their soil will likely result in less need for supplemental minerals.

The minerals known to be essential (up until now, at least) are calcium, phosphorus, sodium, chlorine, iodine, iron, copper, manganese, magnesium, sulfur, zinc, potassium, cobalt, selenium, and molybdenum. Some plants are richer in certain minerals than others, which is a good reason for feeding rations from several different plant sources. The minerals most often lacking are calcium, phosphorus, sodium, and chlorine. Sodium and chlorine, of course, are the constituents of common salt, and salt is always an important element in livestock feeding.

Adding minerals other than sodium and chlorine can be hazardous if not done with caution, knowledge, and good sense. Excesses and imbalances can actually be dangerous.

Moreover, some minerals have definite relationships to others, such as calcium to phosphorus, calcium to zinc, calcium and phosphorus to manganese, and phosphorus to iron. For example, excessive amounts of calcium cause phosphorus manganese, and other elements to be eliminated from the body. Too much potassium can flush sodium from the body. And some minerals make others more effective: copper, for instance, enables the body to utilize iron more fully.

The most important balance concerns calcium and phosphorus. The ration should contain about twice as much calcium as phosphorus. An excess of either interferes with the assimilation of the other, and these two minerals make up about three-fourths of the mineral content of the body.

To demonstrate in one further step the importance of interrelationships, the use of calcium and phosphorus by the body is further dependent upon an adequate supply of vitamin D! What's more, some of the phosphorus in vegetable feeds is not as readily used by the body as inorganic phosphorus or animal phosphorus.

Clearly then, mixing minerals is a job best left to the specialists with knowledge and resources not available to the general farmer, much less the homesteader.

If your land is not yet in good shape, and it's determined that one or more classes of your livestock are in need of additional minerals, commercial mixtures are available. But how can you tell what is needed?

Salt deficiency, for example, may be indicated by unthrifty appearance, loss of appetite, and a marked decrease in milk production (which may go unnoticed except in cows or goats, until it shows up in the young of rabbits, sheep, or other stock). These symptoms are evident because sodium chloride helps maintain osmotic pressure in the body cells, and this pressure is what transfers nutrients to the cells and removes waste matter. In addition, sodium is important in the formation of bile, which aids in the digestion of fats and carbohydrates. Chlorine is required for the formation of hydrochloric acid in the gastric juices, without which proper digestion cannot take place.

For these reasons, additional salt is necessary, even though some is obtained from the feed. Salt is provided in either block or loose form. It is not mixed with the feed, but is provided free choice, since requirements vary from animal to animal, from season to season, according to whether or not milk is being produced (the salt in the milk means the animal needs to ingest more), and other factors.

Although salt is important, animals should not be encouraged to eat an overabundance of it, because a very high consumption of salt can cause other problems, such as flushing other minerals from the body.

Another important mineral which is often not present in sufficient quantity is calcium. Just as with human nutrition, calcium is important for the growth and upkeep of bones and teeth. It is also important in blood coagulation and lactation. It enables the heart, nerves, and muscles to function, and it affects the availability of phosphorus and zinc. Calcium is most often provided by use of ground limestone or oyster shell flour.

Legume forages are important sources of calcium, although even grass hay has more of this mineral than the cereal grains. Legume hay is important for rabbits, sheep, and goats, as it helps minimize calcium deficiencies. However, the hay should be from well-limed fields, there must be a suitable ratio between calcium and phosphorus, and the animal must have sufficient vitamin D either from sunshine or through the ration. Hogs are more likely to suffer from calcium deficiencies than other livestock, especially if fed only cereal grains.

Phosphorus is also required for good bones and teeth and the assimilation of carbohydrates and fats. Phosphorus is a vital ingredient of the proteins in body cells, it's necessary for enzyme activation, it acts as a buffer in blood and tissue, and it occupies a key position in biologic oxidation and reactions requiring energy. Lack of phosphorus can show up in loss of appetite, depraved appetite, lameness, and such problems as milk fever and retained afterbirth, as well as breeding problems. It can cause rickets in young animals, and osteomalacia and related problems in mature animals.

Where both calcium and phosphorus are needed, bone meal is a common supplement. Monosodium phosphate or diammonium phosphate are also used.

Animals deficient in potassium usually have rough coats and have a tendency to chew on wood. (Chewing on wood, however, is normal for rabbits and goats.) Potassium-deficient sheep have dry wool and experience a progressive stiffness from front to rear. However, livestock will seldom show any of these signs, because potassium is usually present in sufficient quantities in roughages.

In the goiter-belt, the Northwest and the Great Lakes region, iodine deficiencies are possible. The enlargement of the thyroid gland (goiter) is the main symptom. When there is not enough iodine in the feed this gland enlarges so that it can produce enough thyroxin, the hormone which controls the rate of body metabolism or heat production. Stabilized iodized salt containing 0.01 percent potassium iodide is the common source of this mineral.

The incorporation of trace minerals in proper amounts in the diets of livestock is highly complicated. It is a good idea to check with a local feed dealer, with local farmers, or with your county agent to determine what minerals are most needed in your area and to work out a balanced diet with them.

Mineral blocks and salt spools come in small sizes designed for use in rabbit cages. The ordinary cattle-sized spools and blocks weigh about 50 pounds each and can be used for smaller livestock as well. Medium-sized blocks and spools are also available in many parts of the country. Salt spools will corrode rabbit cages; they should be hung so that they will not touch the metal cages. Salt spools in pastures and other outside locations should be protected from rain so that pools of brine cannot form. These pools are poisonous to livestock. It is a good idea to protect these blocks and spools from the weather as well. You'll find that they will last a lot longer if they are not eroded by the wind.

Appendix C

VITAMINS

Vitamin deficiencies usually result in mild symptoms which go unnoticed, because rations normally contain enough of the essential vitamins for survival. Top production and condition may be something else. As with minerals, the best way of insuring adequate vitamin intake is by providing feed from properly fertilized soil. And again, since different feeds contain different amounts of different vitamins, a balanced diet is important.

Vitamin deficiencies are most prevalent during periods of drought or when forage is of low quality for other reasons, when animals are in stress, or when production is being forced.

Many goat raisers use injectable ADE (vitamins A, D, and E) almost routinely to insure proper amounts of those important vitamins. Animals produce their own vitamin A out of carotene. Carrots are a good source of carotene, and carrot oil is available commercially. Legume hays contain nine to 14 mg per pound, but the only grain which contains carotene is yellow corn, which has 0.8 to 1.0 mg per pound.

Vitamin D is especially important for pregnant animals, and since stock is commonly kept inside in the winter, when sunlight is less anyway, getting enough of it can be a problem. Sun-cured hay is a good source of vitamin D, with alfalfa containing 300 to 1,000 International Units per pound. Irradiated yeast is the vitamin D source commonly used in commercially prepared feeds.

Vitamin E deficiency can result in white muscle disease

in calves, lambs, and goat kids and increased embryonic mortality in pigs, as well as muscular incoordination in suckling pigs. Vitamin E deficiency seems to be linked with the mineral selenium. While most rations will contain enough of this vitamin, wheat is an especially good source.

B vitamins are seldom a problem with ruminants, as they synthesize them in the rumen. Rabbits practice pseudo-rumination, through coprophagy, or ingestion of certain feces. (See chapter on rabbits.) Green pastures and good hay are sources of most of the B vitamins.

The other vitamins are generally present in common feeds in sufficient quantity and present no major problems.

Appendix D

"COMPANION PLANTING" IN THE BARNYARD

Homesteaders, who frequently learn from observing nature, sometimes ask why certain types of animals can't live together in the barnyard as they do in the wild. In some cases, not only will they tolerate each other but such an arrangement will actually be beneficial, somewhat like companion planting in the organic garden. However, there are certain precautions that should be observed, and common sense should prevail in all situations.

The small poultry flock can fit into a variety of situations outside the standard henhouse. Chickens will act as scavengers, salvaging feed that would otherwise be wasted, thereby producing meat or eggs with relatively little maintenance on the part of the farmer. The consumption of this feed also will help control rodents that would otherwise be attracted to it.

In addition, scratching chickens relish maggots and various other forms of life that aren't exactly welcome in the barnyard. In fact, it caused some amusement among homesteaders recently when an experimental farm operated by a large Midwestern university announced with great pomp that it was experimenting with a "new" means of fly control: scientists let chickens run in the manure piles to destroy the maggots!

Poultry of various kinds have been used in rabbitries to pick up spilled feed, and their scratching under the rabbit cages keeps down insects and promotes more rapid drying of the manure, which helps eliminate odors. However, some rabbit raisers dislike having the aisles littered with chicken manure, and an even more serious objection is that the birds disturb the rabbits. Some breeds of chickens fly more readily than others (Leghorns are quite good fliers, for chickens), and birds on top of the cages are definitely a disturbance. What's worse, the bird droppings then contaminate the rabbit feed, the water, and the rabbits themselves. If chickens are to be kept in the rabbitry they should be of a breed not inclined to fly, and/or the rabbit cages should be hung in such a way that the chickens are prevented from roosting on them. The chickens should have their own roost and nest boxes, as well as drinking water and supplemental feed as necessary.

Some rabbit raisers have reported success using ducks in place of chickens.

Grade A milk regulations prohibit fowl from being in any dairy barn. Since homesteaders are concerned about producing good food, they should logically give serious consideration to following rules at least as stringent as those commercial farmers must observe. The basis for this particular regulation stems from the susceptibility of chickens to tuberculosis, which can be transmitted to cattle, and then to humans. This is a special problem on homesteads since free-ranging birds, and those more than two years old, are the ones most likely to shed pathogenic organisms (germs). (Because commercial poultry flocks are raised in confinement, and because they are not normally kept for more than one year of production, TB in commercially raised poultry has become a relatively minor problem today. On homesteads, the threat is still present, and bears watching.) Extensive soil contamination will cause the disease to persist even if the infected birds are destroyed. TB can also be transmitted to humans directly by fowl.

Ducks and geese are relatively resistant to TB, but pheasants are quite susceptible, especially those raised in captivity. Since many wild birds are carriers, there is no way to absolutely insure a clean flock without total confinement.

Poultry can also infect swine with TB, and therefore the two species should never be allowed together.

Sheep and goats are less susceptible to TB than hogs or cattle, but the disease is not unknown among these animals. It appears to be extremely rare in goats; I have been unable to find a single documented case of TB in goats.

In general, it would seem that caution is required when housing chickens with any other species of livestock. The advantages of having the birds scratch up bedding, helping keep it dry and keeping insect levels down, as well as having them find much of their own food, are too great for most homesteaders to resist entirely. But bc aware of the dangers, immediately remove any sick birds, and keep poultry away from cattle and hogs.

Hogs can be run with cattle with economical results. If whole shelled corn is fed to calves, for example, enough will pass through their digestive systems unutilized to provide one-third of the feed requirements for a 100-pound pig. In other words, with three calves you could keep one hog at no cost except for protein supplement.

With yearling cattle the ratio goes to one pig for every two steers. (But this is with whole shelled corn; the cattle make much more efficient use of ground or rolled grain, and the number of hogs is therefore reduced by half. Four yearling steers on ground or rolled grain will only support one pig. For every 50 bushels of whole corn fed to yearling cattle, approximately 50 pounds of pork can be produced. Since pigs can injure heifers by rooting at the vulva when they're lying down they are usually confined only with steers.

Because of the better digestive system of goats, there is no advantage of running goats and hogs together. Goats are best kept to themselves, for several reasons. If goats are run with cattle, for instance, there are problems with feed and water facilities since the goats can climb into mangers large enough for cattle feeding, and cattle can't eat out of facilities small enough for goats. Cattle are liable to injure goats, too.

Goats kept with horses should have tetanus shots as a precaution. Horses don't *cause* tetanus in goats, but their presence makes goats more likely to contract the disease. Even old horse pastures or barns should be regarded with trepidation by the goatkeeper.

Goats and sheep are both highly susceptible to parasites, and for that reason many farmers frown on keeping them together. The less opportunity for transmission, the fewer worm problems you'll have.

On pasture, however, providing there is adequate room and the right kind of vegetation, sheep, goats, and cattle can work well together. The goats are browsers and will go after tall brush, low trees, and leaves. The cattle eat intermediate growth, and the sheep are very close grazers. But there must be adequate pasture to carry such a population, and pasture rotation takes on added importance to control parasites as well as to allow regrowth.

One of the more unusual forms of symbiosis in the barnyard is the keeping of worms under rabbit hutches. The worms feed on the rabbit droppings and spilled feed, process it into castings which reduces odor and enhances the value of the manure, and they reproduce. The worms can be used in the garden, for fishing, or sold to fishermen and gardeners.

While they may be man's best friend, and few homesteads are complete without a dog, *stray* dogs are never welcome and even trusted family pets can cause trouble. A stray dog just running across your farm can bring in hog cholera, and such dogs have been shot on sight many times by hog farmers. There are always new reports of dogs tearing up even well-built rabbit hutches and killing rabbits by the dozen. Many rabbitries are built with special fences just to keep dogs out. A sheep is defenseless against dogs, and good sheep fencing serves not only to keep sheep in, but dogs out. A sheep torn to shreds by dogs is not a pretty sight. Goats, likewise, are often killed by dogs. Along with the possibility of strangulation, the fact that goats can't run away from dogs when they are tied is one of the most potent arguments against tethering them. And then there are always egg-sucking and chicken stealing dogs.

One interesting point here is that even family pets who have been raised around animals have been known to suddenly go wild and kill one of their "friends."

Dogs also play a role in the cycle of tapeworms. The offal from butchering rabbits and other animals should be buried deeply to prevent dogs from digging it up. Hay, bedding, and feed should never be allowed to become contaminated by dogs.

Appendix E

BARNYARD PESTS

Among the problems of raising livestock are the undesirable pests this activity invites. Foremost among these are flies and rodents.

Flies not only annoy you and your animals but depress the output of your livestock operation. They are filthy disease carriers. They feed in manure. Then, when they sit on your piece of bread, they secrete a fluid which dissolves it so they can digest it. Shoo them off, but you still have manure on your bread. This is in addition to their hairy legs which look so fascinating under a microscope, and which harbor disease carrying germs.

As with most other problems, the best cure is prevention. For flies, this means good sanitation practices. Manure should be cleaned up at least every seven days during the fly season to prevent a completion of their reproduction cycle. Pay special attention to the edges of floors and corners, and around feed bunks or mangers and water tanks. Flies need a moist location to lay their eggs, neither dry nor saturated, and most livestock areas are perfect. Whitewashing or white paint inside barns seems to discourage these pests.

Preventive measures aren't enough, usually, but the organic homesteader is stymied by his reluctance to use chemical sprays. One solution is a homemade fly trap.

This consists of a cylinder of window screen about 17 inches high. It's supported with four or five pieces of wood one-by-two-by-18 inches, with the extra inch forming legs at the bottom. Fashion a top from window screen also. Inside the cylinder, and attached to the top of the one inch leg supports, is a funnel made of window screen. The top is open. The large part of the funnel is at the bottom.

Underneath this place a saucer or pan of rancid meat, sorghum, or similar bait. The flies are attracted to the bait, and when they leave they fly straight up and are funneled into the trap. Such a trap is ecologically sound, harmless to animals and other insects, and catches an awful lot of flies.

Rats and mice can be controlled to some degree by keeping feed in metal containers and by cleaning up spilled feed. Don't provide breeding places, such as piles of junk and lumber. Henhouses and similar buildings can be built off the ground, as open space under them won't provide the dark, damp places these pests need. Obviously, a couple of good cats will do much to keep down the rodent population.

To build a good rat trap, take a large drum. Place it in a position where the little devils will have fairly easy access to it, such as alongside a stack of bags or next to a rough wall.

Keep all other feed cleaned up, but put some feed or grain in the barrel along with a two-by-four that reaches from the top to the bottom so they can climb in and out.

After about a week, when they're bringing all their friends and relatives, remove the two-by-four, put about a foot of water in the drum, and sprinkle the top with bran. I leave the rest to your imagination.

Appendix F

MANURE, THE ADDED BONUS

For the person who starts out as an organic gardener and is never able to get enough compost, the addition of livestock to the homestead yields a pleasing bonus: manure!

It's been said that the total value of all livestock manure produced in the United States was worth almost $5 billion when compared with commercial fertilizer at 1967 prices. In addition, of course, barnyard manure contains organic matter not available in commercial fertilizers, and its nutrients are released more slowly and therefore build up in the soil. Even disregarding these additional benefits, the average farm produces manure worth more than $1,500 a year, which should make it an important "crop."

Yet, much, perhaps even most, is wasted. Some because the farmer simply doesn't realize the value of this by-product, but much more because of lack of information on the proper care of manure or the lack of proper facilities for handling it.

About one-fifth of the nutrients consumed by an animal are passed through the body in the manure. If this manure is used as fertilizer, about one-half is used by the plants in that growing season. Half of the remainder, or one quarter of all the nutrients in the manure is used the following growing season, and half of what's left after that is used the following year. This is in contrast with commercial fertilizers which are, for the most part, gone in the year of application.

Three important points can be drawn from this: In the

first place, it's obvious that a good soil-building program involves maintaining livestock on the farm, properly caring for the manure, and using it as fertilizer. Because of the "half-life" nature of the product, continued applications of manure will continue to improve the soil. One-half the value is gone after the first growing season, but if the same amount is applied the next year and half of *that* fertilizing power is available in addition to what's left from the previous year, and this keeps on building up, in a few years you'll get the same benefits as you would by applying twice as much manure.

On the other hand, since only one-fifth of the nutrients consumed by the animals are returned to the soil in the manure, even if 100 percent of those nutrients were utilized, you're taking five times as much from the soil as you're putting back. Important as it is, manure can't be your *complete* fertilizing program unless you purchase a great deal of feed that's not grown on your land. Other compostable refuse is important too, as well as other forms of organic fertilizers.

The kind and quality of feed affects the quality of the manure to some degree. Higher protein feeds, for example, result in manure higher in nitrogen.

But there are also differences according to species. Sheep manure is famous as a fertilizer. Each ton of pure manure—feces and urine and free from bedding—contains 28 pounds of nitrogen. This compares with 14 pounds for finishing cattle, 13.8 for horses, 11.2 for dairy cows, and 10 for swine. Sheep manure is also high in phosphorus: 20 pounds per ton. On the other hand, sheep produce only about six tons of manure for each 1,000 pounds of body weight, while horses produce eight tons, beef cattle produce 8.5 tons, cows 12 tons, and hogs 16 tons.

Chickens are almost in a class by themselves when it

comes to producing fertilizer. For each 1,000 pounds of body weight, they produce 4.5 tons of manure. But each ton contains 31.2 pounds of nitrogen. It should be used with care on certain crops because the high nitrogen content of the raw uncomposted manure will "burn" them.

If a doe rabbit has 28 young in a year, the family will produce about six cubic feet of manure (without bedding) annually, or about 168 pounds. This is enough for 84 square feet of garden at the recommended application rate of two pounds per square foot.

One advantage of goat manure is that due to the animal's very efficient digestive system, few weed seeds are passed in the manure, a common problem when fresh cattle manure is spread on fields.

Handling Manure

For the subsistence homestead, manure is priceless. Far from being treated as a waste product or as a necessary evil resulting from the production of meat, milk, and eggs, it should be treated with the same care given other valuable homestead commodities.

Manure is a waste product for many commercial farms today. The sheer bulk of it produced in many confinement operations is overwhelming. In many cases, because feed is transported to the animals from other areas, there isn't even land on which to spread manure. In other cases such considerations as a nearby urban population or water pollution dictate that the manure be destroyed rather than conserved and utilized. Its value in comparison with its bulk doesn't permit it to be shipped economically. So manure lagoons, to name just one system, are becoming increasingly common. The manure is conveyed to a series of small ponds which treat it much as a septic tank would.

Since manure handling is becoming a problem of major proportions on modern farms, attempts are being made to eliminate it by eliminating bedding, or as much bedding as possible. Slatted floors are becoming common in hog facilities, and even in cattle barns. The manure goes through slatted floors to pits beneath the building, and bedding and manure hauling are eliminated entirely.

In addition to eliminating a lot of hauling and labor, doing without bedding saves cash. Plant geneticists have come up with wheat and oats that have short straw and are therefore less susceptible to lodging, with the result that the price of straw bedding has increased greatly because of the short supply. In some areas, straw costs as much or more than good hay.

Few of these considerations apply to the homesteader, and even if they did they'd be overwhelmed by one overriding factor: the organic homesteader, by definition, endeavors to live within nature's closed cycle. What has been taken from the soil must be returned. To this end, bedding is important.

Urine makes up 40 percent of the total excrement of hogs; it's less for other animals. But the urine contains nearly 50 percent of the nitrogen, 6 percent of the phosphorus, and 60 percent of the potassium of average manure, or about half of the total fertilizer value. Proper bedding will do much to retain these valuable nutrients.

The finer the bedding the more urine will be trapped and stored. Chopped straw can be twice as absorbent as long straw. Other forms of bedding that help conserve manure are ground corn cobs, sawdust, and wood shavings, and other similar products of local or regional importance.

Even with the important urine salvaged there are still many ways to destroy the value of manure. The most common and the worst is weathering.

The best arrangement I ever had for handling manure was when we lived on a small homestead where the goat barn and henhouse were cleaned twice a year, in spring and fall. This is possible where ceilings are high enough to enable the bedding to be built up for that period of time, and where chickens or other means of controlling flies are employed. The packed down bedding not only stayed dry on top as it got deeper, but it also seemed to give off less odor, and the bottom layers composted where they lay.

Barn cleaning time, to be sure, was a chore, with two to three feet of bedding to haul out! But, coming in the early spring, it enabled me to build compost piles right in the garden, deep enough to prevent rain from seeping through and leaching out the nutrients. With frequent turning, the compost was finished enough to be plowed into the garden by the time we were ready for that operation. In the fall, the barns were again cleaned, the manure stacked in the garden and mixed with after-harvest residues.

This is in contrast with larger farms such as we have now where the cattle and hog pens must be cleaned weekly, if only because of a low ceiling. For a very short time in the spring, and again in the fall, this manure can be spread on the land and immediately disked in. But when there are crops on the land, or there is snow, we have problems, for manure is a highly perishable product and if allowed to dry out or to be weathered and leached, it loses a tremendous amount of its value, especially nitrogen.

We organic farmers and homesteaders know how to save manure, but the more of it we get the harder it becomes to handle properly. The homesteader with a few animals and a few tons of manure can construct a compost bin, but the methods and machines for handling really large amounts of manure just haven't been developed to the same degree. (At least one major company is working on developing these methods and machines now.)

The key conditions for preserving manure are that it must not be allowed to dry out, nor must it be allowed to be leached. The simplest and perhaps best way to conserve it is to apply it directly on the land and work it in before the gasses of fermentation are lost into the air or before natural drainage and leaching wash out the nutrients. In other words, if it's possible to store up six month's worth of manure with a deep litter system, then haul it and incorporate it into the soil immediately, as many of the nutrients as possible would be saved. If manure must be hauled more frequently, it might be possible to use a fallow field, although these are becoming rare. And of course, winter in the North precludes any possibility of immediate incorporation of manure into the earth.

The deep litter system is a possibility for many homesteaders, and may even be best for the animals. Some authorities claim, for example, that chickens raised on deep litter pick up certain antibodies, and in my experience goats on deep litter stay cleaner than those in pens cleaned frequently.

In situations where such a system isn't possible, building a compost heap is advisable. When working with small amounts, the manure and bedding is stacked in a bin. For larger accumulations, it is stacked in long piles. The bins or piles should be deep enough (usually five to six feet is sufficient) so that water will not penetrate them. The piles should not be more than six feet wide to enhance bacterial action. They can be of any length.

Ideally (and here again, it's only practical for small amounts) piles or bins should be covered with plastic. If plastic is not available, a covering of earth will do. They should be formed with a depression to catch rain, because if the material gets too dry, it will lose much of its nitrogen and some of the organic matter will escape into the air as carbon dioxide.

Such a pile can then be considered secure, and applied to the land when conditions permit. In contrast, manure carelessly stacked in the barnyard, or spread on the fields and allowed to dry out or leach out in rain, is really and truly a "waste."

Amounts and Fertilizing Value of Some Animal Manures

Animal	Tons excreted per year per 1,000 lbs. live-weight	% Nitrogen	% Phosphoric acid	% Potash
Rabbit	4.2	2.4	1.4	0.6
Sheep and goat	6.0	1.44	0.5	1.21
Swine	16.0	0.49	0.34	0.47
Chicken	4.5	1.0	0.8	0.39
Dairy cow	12.0	0.57	0.23	0.62
Beef steer	8.5	0.73	0.48	0.55
Horse	8	0.70	0.25	0.77

Appendix G

BREEDING TABLES

Animal	*Age of puberty (months)*	*Interval of heat (days)*	*Average duration of heat (hours)*	*Average gestation period (days)*
Rabbits	*6- 8*	*—not applicable—*		*31*
Swine	*4- 7*	*18-24*	*2- 3 days*	*114*
Sheep	*5- 7*	*14-20*	*30*	*148*
Goats	*4- 8*	*12-25*	*36-48*	*151*
Cattle	*8-12*	*21*	*16-20*	*283*
Horses	*12-15*	*21*	*4- 6 days*	*336*

Appendix H

WATER ABSORPTION OF VARIOUS BEDDING MATERIAL

Material	*Lbs. of water absorbed per cwt. dry bedding*
Peat moss	*1,000*
Chopped oat straw	*375*
Oat straw long	*280*
Vermiculite	*350*
Wood chips (pine)	*300*
Hardwood wood chips	*150*
Wheat straw, chopped	*295*
Wheat straw, long	*220*
Sawdust (pine)	*250*
Sawdust (hardwood)	*150*
Corn stalks, shredded	*250*
Peanut hulls	*250*
Sugar cane bagasse	*220*
Corncobs, ground	*210*
Broadleaf leaves	*200*
Sand	*25*

Appendix I

SOME HELPFUL MEASUREMENTS

*To estimate the number of tons of hay in a stack, mul tiply the length by the width in feet. Multiply that by one-half the height. Divide by 300, and you'll have the answer in tons.

*A bushel contains 2,150.42 cubic inches.

To find the number of bushels of grain in a bin, simply take four-fifths of the volume in cubic feet. Example: A bin 10 feet long and five feet wide has two feet of grain stored in it. $10 \times 5 = 50 \times 2 = 100 \times \frac{4}{5} = 80$ bushels.

*If corn is stored on the cob, multiply the number of cubic feet in the crib by four and divide by ten.

*An acre is 43,560 square feet, or 4,840 square yards.

*A square acre measures 208.71 feet on each side.

**Number of pounds to the bushel*

Alfalfa seed	*60*
Barley	*48*
Soy Beans	*60*
Bran	*20*
Buckwheat	*48*
Shelled corn	*56*
Ear corn	*70*
Oats	*32*
Wheat	*60*

Appendix J

CURING AND SMOKING MEATS

When we think of cured, smoked meat, we generally think of ham and bacon. And the homestead pork producer is certainly missing a taste thrill if he chooses to forego the extra labor and time involved in curing and smoking his home-produced pork. The home-cured product compares to store bought ham and bacon at about the same ratio homemade bread compares to the mushy goo that passes for bread in grocery stores. You have probably never tasted bacon like it in your life.

But did you know you can also cure and smoke (or cure *or* smoke) just about anything else? Smoked pheasant is a delicacy almost everyone has heard of even if most people haven't tasted it, but if you don't raise pheasants why not try a guinea or a whole rabbit? We've used a backyard smokehouse to smoke carp, a rough fish no self-respecting middle-class person in this part of the country would stoop to eat . . . and it was delicious. Even some cheeses can be enhanced by the flavor of hickory.

Any meat can be cured without being smoked, or smoked without being cured. If you do both, as with ham and bacon, the curing comes first.

Just as any good cook has her or his own recipes for special dishes, there are hundreds of recipes for curing brines. Here's one that works for me:

It requires eight pounds of salt, two pounds of sugar, two ounces of saltpeter and four and one-half gallons of water. Be sure the salt has completely dissolved before putting the meat into the pickle.

Put the meat into a crock or plastic container (salt corrodes metal) and pour the brine over it. A weight, such as a plate or platter with a gallon jug of water on it, will hold the meat under the solution.

Meat should be left in the pickle for about four days for each pound. In other words, a 15-pound ham takes about 60 days. To prevent spoilage and to insure that all parts of the meat are cured uniformly, take the meat out every seven days and repack it.

The solution should be kept fairly cool. However, if it does get slimey, no harm. Take the meat out and wipe it down with a cloth and cold water, mix another batch of brine, and keep on the original schedule.

A very simple, temporary smokehouse, can be constructed by digging a pit about two feet square and two to three feet deep. Dig a six-by-six-inch trench away from this pit, uphill if possible, about 10 to 12 feet long.

The pit is where you build your fire. The trench is the flue that carries the smoke to the meat. Cover the trench with a one-by-10-inch board, and heap earth on top of that. For a more permanent installation, consider using some form of tile.

At the end of the trench opposite from the firepit, place a barrel. Wooden barrels are hard to come by these day, but 50-gallon drums work just as well. In fact, one of my first smokehouses of this type was made from a large electrical outlet box that happened to be lying around. Use whatever you have.

Remove both ends of the barrel, drill two holes on opposite sides near the top, place a broom handle in the two holes, and put a wooden cover over the top of the barrel. The meat is hung by wires from the broom handle.

Start a fire in the pit. Hickory and apple are ideal woods to use, although sections of the country where these aren't available have local substitutes. Hardwoods are best; avoid pitchy woods such as pine.

When the fire has a good start, it might be necessary to smother it with hardwood sawdust. You don't want a hot fire, but a steady, cool one. Cover the firepit with a sheet of tin (although I've already used scrap plywood with only minimal scorching) and bank the whole thing with earth so the smoke will be drawn through the tunnel and thus into the smoke chamber. You can get a lot fancier, of course, especially with permanent installations, but the principles are the same.

With hickory, two sides of bacon are just the way we like them in 24 hours. Since there is little draft, two or three small pieces of hickory provide all the smoke needed in this period.

The actual time for any meat depends not only on your personal taste, but on the construction of the smokehouse, the intensity of the fire, and any other factor affecting the amount of smoke reaching the meat. You must avoid "cooking" the meat (which is why I prefer the trench-type smoker rather than the simpler method of merely setting the barrel over the fire). Too-heavy smoke overemphasizes the smoke flavor. Total smoking time can range from a few hours to a few days.

For more information on curing and smoking meats, I refer you to the following material:

"A Complete Guide to Home Meat Curing." This is available from Morton Salt Company, P.O. Box 355, Argo, Illinois 60501.

"Slaughtering, Cutting, and Processing Pork on the Farm." Farmer's Bulletin No. 2138, U.S. Department of Agriculture. Available from the Superintendent of Documents at the U.S. Government Printing Office, Washington, D.C. 20402. (State Agricultural Extension Services have similar pamphlets available.)

Stocking Up: How to Preserve the Foods You Grow, Naturally, by the editors of *Organic Gardening and Farming* magazine. Available from Rodale Press, Emmaus, Pennsylvania 18049.

Homebook of Smoke-Cooking Meat, Fish, and Game, by Jack Sleight and Raymond Hull. Available from Stackpole Books, Harrisburg, Pennsylvania 17105.

Appendix K

TANNING

There is very little commercial demand for the pelts from homestead livestock. With the prices paid, the small quantity produced, the effort involved in making raw pelts saleable, and shipping costs, it's generally more economical to use the skins on the compost heap rather than sell them. They are high in both nitrogen and phosphorus.

But this doesn't mean that pelts from the homestead herds are useless. On the contrary, anyone who's so inclined can easily do an acceptable job of tanning those skins for use in a variety of craft projects. Simple items such as mittens, hats, purses, and more complex ones such as vests and even jackets, will turn "waste" pelts into valuable by-products.

Tanning can be a quite complicated and tedious chore. I've used tanning solutions that called for chemicals the druggist had to order special for me, and have run across other formulas that use chemicals even the pharmacist claims he never heard of.

But here's a method that has been used by many homesteaders with great success, and which calls for ingredients that are readily available.

While this solution can be used for any type of fur, let's talk about rabbit skins. Not only are rabbit furs the most common on the homestead: the texture of good ones makes them useful and beautiful, the wide variety of colors and color patterns are interesting to work with, and perhaps most important for a first-time tanner, you'll feel more at ease with a rabbit skin than with something more ponderous. Save the awesome steer hide for later.

The tanning agent in this method is sulfuric acid. To make it even simpler, use battery acid, available from any garage or auto supply store. Battery acid is dilute sulfuric acid.

Here's the recipe:

Two ounces of sulfuric acid *or* eight ounces of battery acid

Two pounds of salt—any cheap kind.

A three- to five-gallon crock or similar non-metallic container. A plastic garbage pail will do.

Two gallons of water.

A weight to hold down the skins in the solution: A glass jug filled with water, a scrubbed brick or rock, or anything similar.

Add the salt to the water. Then tip your container and let the acid dribble down the side into the water. Never add water to acid, and be careful not to let it splash. Stir the solution with a wooden stick. At this point the acid is dilute enough that it's quite safe, even if it touches your skin.

Keep it at 70° F, or as close to it as possible. Higher temperatures can damage pelts, and lower temperatures retard the tanning process.

Now you're ready to tan.

To be sold, pelts must be fleshed, carefully stretched, and properly dried. Contrary to popular opinion, the hides are not salted. But these time-consuming steps are eliminated with home tanning.

We usually rinse the skins in a bucket of cold water to remove any blood stains. Adding about two cups of salt to a gallon of water seems to aid in the fleshing process we'll come to later, according to some people. Then wash the skin in warm water and detergent, and squeeze out the excess water. Never wring a fur: squeeze it.

Finally, toss the hide into the tanning solution (be sure the salt has dissolved) swish it around a bit with the wooden stick, and weight it down to keep it from floating.

A small hide, in full-strength tanning solution at 70° F. will be finished in about three days. Naturally the solution grows weaker as more pelts are tanned, larger pelts take somewhat longer, and lower temperatures slow down the process. You can adjust to these variables with a little experience: they're nothing serious. The pelts can stay in the solution for up to a year, in fact, as long as they're stirred from time to time.

When a skin is ready it's taken out, washed in detergent and rinsed in cool water. At this point the fat and flesh should separate from the hide with only mild resistance. If it's really prime and you're very careful, you can separate the flesh from the hide in a single piece. This is a far cry from the labor of fleshing a fresh hide!

After fleshing, wash and rinse again and return to the tanning vat for another week, or longer.

Then run it through a final wash-rinse-squeeze process. Hang it in a shady place to drip dry, not in the sun. While it's still damp and limp, put it in the clothes dryer. If it's too wet, it won't tumble properly. Leave the heat off: you're not drying the fur, just tumbling it.

In fact, if a dryer isn't available, this step can be eliminated. It just makes the next step easier. And that's "breaking" the skin. Breaking is gently pulling and stretching small areas of the skin in different directions. The stiff, brown hide will turn white and soft.

And that's all there is to it.

Oh yes, as one homesteader told us, there's a drawback to this that isn't mentioned in the books on tanning. Once word of your talent gets around, every dead critter within 25 miles will wind up on your doorstep. I once turned down the opportunity of tanning the beautiful fur . . . of a skunk that had been killed by a car.

Appendix L

MAKING SOAP

Two of the most important homestead axioms are: nothing is wasted; and we provide as many of our own necessities as possible. Therefore, one of the chores of the homestead livestock producer is making soap.

Hog butchering is the most common occasion for soapmaking, since hogs have by far the most fat. But soap can be made from any animal fat. Poultry fat produces a very soft soap, objectionably soft. Goat tallow makes a very hard soap.

Even after rendering lard from the pig for cooking and baking, there are still many pieces of fat undesirable for rendering that will produce good soap. Homemade soap obviously contains no phosphates or detergents: you can water plants with the dishwater if you want to. In fact, this is the kind of soap used for washing dishes back in the days when pigs were swilled with dishwater. Don't try that with your store-bought kind!

Chances are you can buy a can of lye that will have a basic soap recipe printed right on it. But there is a great deal of room for experimentation and creativity, and you'll probably want to try making several different kinds.

Even with the basic recipe you can add a pine aroma by boiling pine needles slightly in soft water and using the pine water in the soap making process. Or you can perfume the soap with flowers, if that sort of thing appeals to you. Press any strong smelling flowers into the lard for 24 hours before making your soap.

Then place 10 pounds of lard in a kettle with two quarts of water. Bring to a boil, and set it aside to cool overnight. Any dirt and meat particles will settle out.

Outdoors, add two quarts of soft water to two cans of lye. Use a stoneware crock or other non-metallic vessel and be sure to use cold water or the lye will fume. In fact, the cold water will become quite hot from the lye. Allow it to cool.

Then add four tablespoons of sugar, two tablespoons of salt, six tablespoons of powdered borax and one-half cup of ammonia to one cup of soft water. Pour it into the cooled lye solution. Add the cooled lard, and stir well with a wooden paddle.

When the soap is somewhat mushy, we pour it about one inch thick into cut-down cardboard boxes lined with newspaper. Don't pour it too thick, as that will result in thick bars that are hard to handle when you use them. Before the soap is completely hard, cut it into bars of the desired size. You can also mold it into various shapes at this point if you want to get fancy about it.

The longer the soap ages, the harder it gets. Our best product has been about a year old, although you can use it as soon as it's hardened, of course.

Appendix M

GOING BIGTIME

Invariably, the day arrives when the fledgling goatkeeper or rabbit raiser considers getting larger and blossoming into a full-time operation. Many of them write to magazine editors for advice, and the advice I provide is quite simple: don't *go* into business: *grow* into business.

There are no hard and fast rules, but in general a love of the animals and a good working knowledge of their care and feeding, important as they are, simply aren't enough to insure success. What is required is much more capital than most people are willing or able to invest, and even more important, much more knowledge of and attention to business details than most beginning farmers can imagine.

If you decide to raise pigs or beef cattle or to produce cow milk, getting rid of your output is no problem. There are established market channels for such products. All you have to do is to make sure your income exceeds your outgo, which is management.

But rabbits, goats, and some of the other more exotic agrarian pursuits are quite different. In most places there are no established, stable markets. That means that in addition to raising rabbit fryers, for example, you'll have to do the slaughtering and dressing . . . and the selling as well. At this point you find yourself a full-time butcher or salesman, when all you wanted to do was raise a few rabbits to sell.

The prospects for goat dairying are even more disheartening. While recently enacted federal legislation governing rabbit meat inspection doesn't include very small rabbitries, laws covering milk are much more stringent. In most cases, the catch is that sales of raw milk are illegal, and yet pasteurizing a few gallons, or even a few hundred gallons, is totally impractical from a cost standpoint.

Even more interesting is the fact that where raw goat milk sales are legal, and where serious advertising campaigns have been carried out, there still is no big rush to goat milk. The demand simply isn't there.

One mistake beginners make, especially in the field of goat dairying, is assuming that they can beat the price of established goat dairies. There aren't many of them, and their prices are completely in line with their costs. The fellow who tries to underprice them soon finds himself in hock because he didn't consider all the costs of producing the milk.

This isn't to say there aren't opportunities for full-time employment in the field of small stock. There are several areas where markets for goat milk for either cheese or fluid milk exist. There are even more where reputable rabbit processors will take all you can raise. In all cases that I know of, the price paid is fair but not extravagant, and it will take much experience and attention to details to make the operation pay.

In addition, there are other areas where the lone operator can utilize ethnic tastes or his own special talents to make a success of his operation.

In other words, asking "Can I be a successful rabbit (or goat or whatever) farmer?" is a loaded question. I don't know, any more than I would know if you could be a successful shoe salesman or lawyer.

One thing is certain. Stay away from anybody who guarantees that you'll make $10,000 a year raising rabbits or guinea pigs. It's so easy to get into small stock raising, and so many people are interested in the field, that it's a natural spawning ground for promoters and fast-buck operators. Don't sink your life's savings into a venture until you have made darn sure, by modern business standards, that a market exists and that you can handle both the production and marketing details.

Goat milk and rabbit meat are not on every American table for a very good reason: they cost more, because they cost more to produce. The urbanite can't afford them, and doesn't care when cheaper substitutes are available.

The homesteader, on the other hand, finds these "high-cost" food items cheaper than the usual products for reasons we've discussed in the last 60,000 words or so.

There are commercial possibilities, to be sure. But you wouldn't go into the restaurant business, or start a hardware store or anything else, without adequate background and financing. Just be certain you have the background before diving into a small stock enterprise.

Appendix N

WHERE TO GO FOR HELP AND MORE INFORMATION

There are numerous livestock associations whose purpose is to promote their breed. This naturally entails helping beginners find good stock, providing help and information so they can improve their stock and their methods, and of course the associations handle registrations, shows, have annual conventions, and so on.

There are two national goat clubs. The American Dairy Goat Association (ADGA) is located at Spindale, North Carolina, and the American Goat Society (AGS) has its office at 1606 Colorado St., Manhattan, Kansas 61502. In addition there are clubs promoting specific breeds, but since their secretaries are elected on an annual basis and therefore the addresses change, it would be advisable to contact ADGA for the current address of the club you're interested in. Most have newsletters that list sources of stock and stud service, as well as providing helpful information. There are also state and local clubs in many areas. Some are quite active, others are not, but it will pay to check them out.

The national rabbit club is the American Rabbit Breeders' Association, 1007 Morrisey Dr., Bloomington, Illinois 61701. The membership fee includes the magazine, *Domestic Rabbits*. There are also specialty clubs for many of the most popular breeds of rabbits and again, since the addresses change frequently, it would be well to ask the secretary of ARBA about the specific breed you're interested in. Most of these clubs issue guide books covering their breed, and have newsletters. And, as with goats, there are state and local associations whose addresses are available from the national secretary.

These organizations can provide help and information that the more general sources of agricultural information might not be aware of. But these general sources can also provide a great deal of help in certain cases. The most useful one is probably the county agent, located in the county seat.

Information is available from the Department of Animal Science of each state university. Each state also has a State Department of Agriculture, usually located in the state capital. (Exceptions: Maryland, Nevada, and New Mexico.) These departments offer many services to raisers of small stock, often including diagnostic work.

The U.S. Department of Agriculture, Washington, D.C. 20505, has a large staff of specialists and many publications. For a list of the publications, write to Office of Information, U.S. Department of Agriculture, Washington, D.C. 20505. Many of these are available through the county agent or from your senator or congressman.

Don't forget your local agriculture instructor. Future Farmers of America (FFA) is an organization of boys enrolled in vo-ag. 4-H clubs are sponsored by the agriculture extension service of the land grant colleges. For information contact your county agent.

Many commercial feed companies have information on small stock. Carnation-Albers, 6400 Glenwood, Suite 300, Shawnee Mission, Kansas 66202, has some excellent material on rabbits; and Ralston Purina Company, Checkerboard Square, St. Louis, Missouri 63188, has material on both rabbits and goats. Many smaller companies also provide literature.

Appendix O

FOR FURTHER READING

Magazines

Countryside & Small Stock Journal—Jerome D. Belanger, editor and publisher, Rt. 1, Box 239, Waterloo, Wisconsin 53594. $5 per year. Often called "the *Organic Gardening & Farming* of small stock." Heavy on rabbits, goats, and poultry.

Dairy Goat Journal—Kent Leach, editor and publisher, Box 1908, Scottsdale, Arizona 85252. $4 per year. Official publication of the American Dairy Goat Association. A good buyer's guide.

Domestic Rabbits—Robert W. Bennett, editor. Published by the American Rabbit Breeders Association and furnished as part of the membership benefits. Membership is $5 per year. ARBA, 1007 Morrissey Dr., Bloomington, Illinois 61701.

Books

Goats:

Aids to Goatkeeping, by Corl Leach. $6. This slim volume has been around a long time, but it's still one of the "must" books for the goat raiser. Available from *Dairy Goat Journal.*

Best of Capri, Elsie Evelsizer, Judy Kapture, and Jerome Belanger, editors. $3. A collection of information from goat club newsletters.

Goat Husbandry, by David McKenzie. $16.75. The Bible of goat raising. Expensive, but worth it to the serious breeder.

Starting Right with Milk Goats, by Helen Walsh. $3. Reprinted from the famous old "Have-More Plan." Probably the best source of basic information for beginners.

(The last three are available from *Countryside & Small Stock Journal.*)

Rabbits:

Domestic Rabbit Production, by George Templeton, former Director of the U.S. Rabbit Experiment Station at Fontana, California. $7.95. The most comprehensive, authoritative book on rabbits. Available from *Countryside & Small Stock Journal.*

How to Start a Commercial Rabbitry, by Paul Mannell. $2.95. A good source of information on raising rabbits for meat—on the homestead or on a commercial scale. Available from *Countryside & Small Stock Journal.*

Poultry:

Starting Right with Poultry, by G. T. Klein. $3. This book from the "Have-More Plan" series is the only one we've found that recognizes that some people raise just a few chickens in their backyards. Recently reprinted by Garden Way Publishing, Charlotte, Vermont 05445, it's also available from *Countryside & Small Stock Journal.*

INDEX